Statistics 1

Greg Attwood
Gordon Skipworth
Gill Dyer

Published by Pearson Education Limited, a company incorporated in England and Wales, having its registered office at Edinburgh Gate, Harlow, Essex, CM20 2JE. Registered company number: 872828

Edexcel is a registered trademark of Edexcel Limited

Text © Greg Attwood, Gordon Skipworth and Gill Dyer 2009

First published 2001 under the title *Heinemann Modular Mathematics for Edexcel AS and A level: Revise for Statistics 1*

12 11
10 9 8 7 6 5 4 3

British Library Cataloguing in Publication Data is available from the British Library on request.

ISBN 978 0 435519 30 8

Copyright notice
All rights reserved. No part of this publication may be reproduced in any form or by any means (including photocopying or storing it in any medium by electronic means and whether or not transiently or incidentally to some other use of this publication) without the written permission of the copyright owner, except in accordance with the provisions of the Copyright, Designs and Patents Act 1988 or under the terms of a licence issued by the Copyright Licensing Agency, Saffron House, 6–10 Kirby Street, London EC1N 8TS (www.cla.co.uk). Applications for the copyright owner's written permission should be addressed to the publisher.

Edited by Susan Gardner
Typeset by Techset
Illustrated by Techset
Cover design by Christopher Howson
Cover photo/illustration © Edexcel
Printed in Malaysia, CTP-KHL

Acknowledgements
Every effort has been made to contact copyright holders of material reproduced in this book. Any omissions will be rectified in subsequent printings if notice is given to the publishers.

Disclaimer
This Edexcel publication offers high-quality support for the delivery of Edexcel qualifications.

Edexcel endorsement does not mean that this material is essential to achieve any Edexcel qualification, nor does it mean that this is the only suitable material available to support any Edexcel qualification. No endorsed material will be used verbatim in setting any Edexcel examination/assessment and any resource lists produced by Edexcel shall include this and other appropriate texts.

Copies of official specifications for all Edexcel qualifications may be found on the Edexcel website – www.edexcel.com

About this book

This book is designed to help you get your best possible grade in your Statistics 1 examination. The authors are senior examiners.

Revise for Statistics 1 covers the key topics that are tested in the Statistics 1 examination paper. You can use this book to help you revise at the end of your course, or you can use it throughout your course alongside the course textbook, *Edexcel AS and A-level Modular Mathematics Statistics 1*, which provides complete coverage of the specification.

Helping you prepare for your examination

To help you prepare, each topic offers you:

- **What you should know** – a summary of the statistical ideas you need to know and be able to use.
- **Test yourself questions** – help you see where you need extra revision and practice. If you do need extra help they show you where to look in the *Edexcel AS and A-level Modular Mathematics Statistics 1* textbook.
- **Worked examples and examination questions** – help you understand and remember important methods, and show you how to set out your answers clearly.
- **Revision exercises** – help you practise using these important methods to solve problems. Examination-level questions are included so you can be sure that you are reaching the right standard, and answers are given at the back of the book so that you can assess your progress.

Examination practice and advice on revising

Examination style paper – this paper at the end of the book provides a set of questions of examination standard. It gives you an opportunity to practise taking a complete examination before you meet the real thing. The answers are given at the back of the book.

How to revise – for advice on revising before the examination, read the How to revise section on the next page.

How to revise using this book

Making the best use of your revision time

The topics in this book have been arranged in a logical sequence so you can work your way through them from beginning to end. However, **how** you work on them depends on how much time there is between now and your examination.

If you have plenty of time before the examination then you can **work through each topic in turn**, covering the key points and worked examples before doing the revision exercises and test yourself questions.

If you are short of time then you can **work through the Test yourself sections** first, to help you see which topics you need to do further work on.

However much time you have to revise, make sure you break your revision into short blocks of about 40 minutes, separated by five- or ten-minute breaks. Nobody can study effectively for hours without a break.

Using the Test yourself sections

Each Test yourself section provides a set of key questions. Try each question.

- If you can do it and get the correct answer then move on to the next topic. Come back to this topic later to consolidate your knowledge and understanding by working through the what you should know section, worked examples and revision exercises.
- If you cannot do the question, or get an incorrect answer or part answer, then work through the what you should know section, worked examples and revision exercises before trying the Test yourself questions again. If you need more help, the cross-references beside each Test yourself question show you where to find relevant information in the *Edexcel AS and A-level Modular Mathematics Statistics 1* textbook.

Reviewing what you should know

Most of what you should know are straightforward ideas that you can learn: try to understand each one. Imagine explaining each idea to a friend in your own words, and say it out loud as you do so. This is a better way of making the ideas stick than just reading them silently from the page.

As you work through the book, remember to go back over what you should know sections from earlier topics at least once a week. This will help you to remember them in the examination.

Mathematical models in probability and statistics 1

What you should know

1 **Mathematical model** – a simplified mathematical version of a problem devised to describe, or make predictions about, a real-world situation.

2 **Prediction** – a result or outcome predicted by a mathematical model.

3 **Observed outcome** – results of observations of the real world e.g. experimental data.

4 **Statistical test** – a test used to compare a prediction with an observed outcome.

5 **Refining the model** – reformulating as a new model which might make predictions closer to the observed outcomes.

6 **Steps in modelling** –
1. Devise a model.
2. Use it to make predictions.
3. Collect experimental data from the real world.
4. Compare predictions with collected data and test statistically to see how well the model describes the real world.
5. If necessary refine the model.
6. When satisfied accept the model.

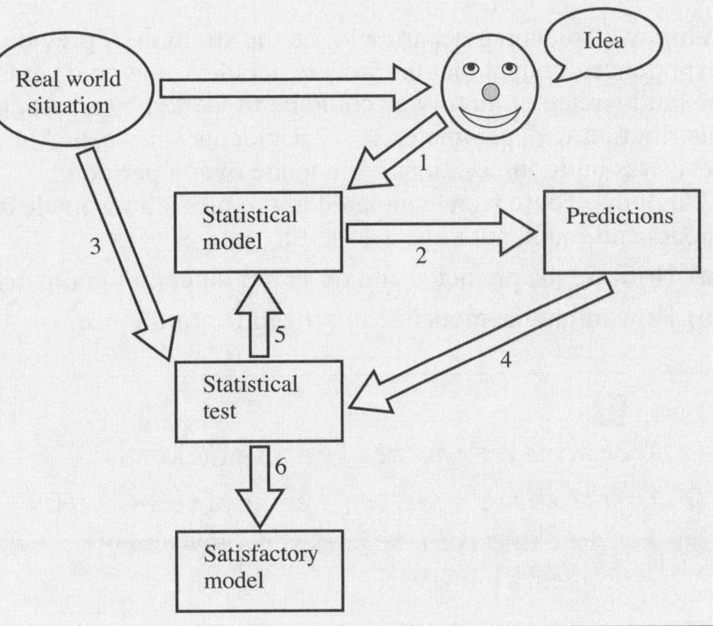

Test yourself

What to review

Review Edexcel Book S1 pages 1–3

This chapter is short and yet it contains important ideas that lie behind much of statistics. If you are confident that you can answer the two questions on page 3 then do so and check your answer against those given at the back of the book. If you have any doubts read through this section before attempting them.

Example 1

(a) Explain what you understand by a statistical model.

(b) Write down a random variable that could be modelled by:

(i) a discrete distribution

(ii) a normal distribution.

Using **1**,

(a) A statistical process devised to describe or make predictions about the expected behaviour of a real-world problem.

(b) **(i)** The number uppermost on a die after it has been rolled.

(ii) The height of adult males.

Example 2

A highway planning department, on the strength of previous experience, thought that the weekly accident rate on a yet to be built stretch of motorway could be modelled by a particular distribution with parameter $\lambda = 2$ accidents per week. After the road was built observations were made over a period of 12 months. These were compared to the predictions made by the model and found not to be a good fit.

(a) How are the predicted and observed outcomes compared?

(b) How might the model be improved for future use?

Using **4**,

(a) They would be compared by statistical tests.

(b) To improve the model the average observed weekly accident rate could be used for λ instead of 2 accidents per week.

Revision exercise 1

Throughout statistics, there are references to the modelling process. In *Statistics S1* modelling is specifically referred to in chapter 9 – The normal distribution.

1 Explain the process of statistical modelling.

2 Name three real-world problems that might be modelled by a normal distribution

Representation and summary of data – location

2

What you should know

1 **Data** – a series of observations, measurements or facts.

2 **Variable** – that which is measured or observed, e.g. height.

3 **Type of variable**
Quantitative variable – one with measurements given as numbers.
Qualitative variable – one to which numbers cannot be assigned.
Discrete variable – one taking only specified values.
Continuous variable – one taking any value within a given range.

4 **Frequency distribution** – shows the values of a variable and how often each occurred.

5 **Cumulative frequency** – obtained by adding frequencies one at a time across a row of frequencies.

e.g.				
frequency	1	4	8	…
cumulative frequency	1	5	13	…

6 **Grouped frequency distribution** – a table in which frequencies are associated with groups or classes rather than single observations.

7 **Terminology** – consider the class 8–10 shown here.

class width
lower class boundary 7.5 8 – 10 10.5 upper class boundary
9
lower class limit
class mid-point
upper class limit

8 **Mode** – the value of a variable that occurs most frequently.

9 **Median** – the middle value of an ordered set of data.

10 **Mean** – the sum of all observations divided by the total number of observations.

$$\text{Population mean} = \mu = \frac{\sum x}{n} \text{ or } \frac{\sum fx}{\sum f}$$

$$\text{Sample mean} = \bar{x} = \frac{\sum x}{n} \text{ or } \frac{\sum fx}{\sum f}$$

11 **Coding** – transforms a variable into a simpler one and makes for easier arithmetic,

e.g. $y = \frac{x - a}{b} \Rightarrow \bar{x} = b\bar{y} + a$

> You need to know how to calculate the mode, median and mean; however, questions in the S1 examination will be directed more towards interpretation than calculation.

Test yourself

1 **(a)** For the data in the table show that $\sum fy = -132$, where $y = \frac{x - 384.5}{5}$ and x and f are the mid-point and the corresponding frequency for each of the classes.

Lifetime	350–359	360–369	370–374	375–379
Number of bulbs	3	21	38	62

Lifetime	380–389	390–399	400–429
Number of bulbs	75	37	14

(b) Calculate $\bar{y}$ and hence find an estimate of the mean lifetime of the bulbs.

What to review

If your answer is incorrect:

Review Edexcel Book S1 pages 24–26

Example 1

For each of the following variables state whether they are discrete or continuous.

(a) The number of children in a class

(b) The time spent reading

(c) The number of days in a month.

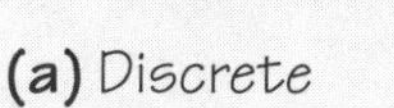

(a) Discrete (b) Continuous (c) Discrete

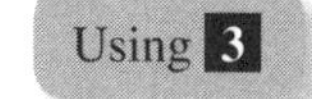

Using **3**

Example 2

For each of the following classes write down the class boundaries, mid-point and class width.

(a) 1–5 **(b)** 3.5–4.4 **(c)** −4–(+4).

(a) 0.5, 5.5; $\frac{1}{2}(0.5+5.5) = 3$; $5.5 - 0.5 = 5$

(b) 3.45, 4.45; 3.95; 1.0

(c) −4.5, 4.5; 0; 9

Using **7**

Example 3

The following observations are the number of printing errors found on the first thirty pages of a book.

2	1	2	5	4	1	2	4	1	3
3	4	0	1	2	5	1	3	1	2
6	1	1	4	3	2	1	6	2	5

(a) Summarise these data using a frequency distribution. Include both frequencies and cumulative frequencies.

(b) Find the percentage of pages containing fewer than three errors.

(a)

Error	Tally	Frequency	Cumulative frequency
0	|	1	1
1	~~||||~~ ||||	9	10
2	~~||||~~ ||	7	17
3	||||	4	21
4	||||	4	25
5	|||	3	28
6	||	2	30
		Total 30	

Using **4** and **5**

(b) From the cumulative frequency column 17 pages had fewer than 3 errors.

$\therefore$ Percentage $= \frac{17}{30} \times 100 = 56.7\,\%$

Example 4

Ten randomly chosen children were each asked how many children there were in their family. The results are as follows:

1 3 4 5 2 3 2 2 4 2

Find:

(a) the mode **(b)** the median

(c) the mean of these data.

Ordering the data gives: 1 2 2 2 2 3 3 4 4 5

(a) Using **8**, Mode = 2

(b) Using **9**,

$$\text{Median} = \tfrac{1}{2}(\text{5th} + \text{6th}) \text{ observations} = \tfrac{1}{2}(2 + 3) = 2.5$$

(c) Using **10**, $\text{Mean} = \dfrac{1 + 2 + \ldots + 5}{10} = \dfrac{28}{10} = 2.8$

Remember to order the data.

Note that neither the median nor the mean has to be a whole number.

Worked examination question 1 [E]

A sixth former examined the books returned by 100 people using the library one Saturday. She calculated that the mean number of times these books had been borrowed over the last 12 months was 8.2. State giving a reason, whether you would expect the mean number for all the books in the library to be higher or lower than 8.2.

It is expected that the mean would be lower than 8.2. The sample taken is biased, since it is based on books which have actually been borrowed; taking into account the books which are seldom or never borrowed would produce a lower result for the true value of the mean.

Revision exercise 2

1 A shop buys mugs from a pottery in batches of 50. When each batch is delivered the shop manager checks them for defects. The number of defective mugs in each of 30 batches is shown below.

6 9 8 1 2 4 8 6 4 7
2 0 2 6 4 1 0 7 7 0
6 5 3 2 3 3 3 1 1 1

(a) Construct a frequency distribution to summarise these data. Include both frequencies and cumulative frequencies.

The manager has decided to buy from a different pottery if more than 10 batches have 6 or more defective mugs in them.

(b) Explain whether or not the manager will have to buy from another pottery.

2 A secretary recorded the number of photocopies she made each day during a particular month. The results are shown below.

Number of photocopies	10	11	12	13	14	15	16	17	18
Number of days	2	3	3	7	6	4	3	1	1

(a) Write down the modal number of photocopies made per day during that month.

(b) Find the median number of photocopies made per day.

(c) Calculate the mean number of photocopies made per day.

(d) Which of these three measures should she use to estimate the number of photocopies she makes in a year?

Representation and summary of data – measures of dispersion

3

What you should know

1 **Range** – the value obtained when the smallest observation in a data set is subtracted from the largest one.

Range and IQR will not be the direct focus of questions in the examination paper. They may be used to draw inferences or to help in interpretation.

2 **Quartiles** – divide the data into four equal parts:

- 25% of the observations are less than or equal to the first quartile Q_1
- 50% of the observations are less than or equal to the second quartile (median) Q_2
- 75% of the observations are less than or equal to the third quartile Q_3.

3 **Interquartile range (IQR)** – the value obtained when the lower quartile is subtracted from the upper quartile.

$\text{IQR} = Q_3 - Q_1$

4 **Semi-interquartile range (SIQR)** – half the interquartile range.

$\text{SIQR} = \frac{1}{2}(Q_3 - Q_1)$

5 **Percentiles** – divide the data into 100 equal parts.

6 **Population** – a collection of individual items or individuals.

7 **Sample** – a subset of a population used to represent that population.

8 **Variance** – the variance of a population of observations $x_1, x_2, \ldots, x_n$ is the mean of the sum of the squared deviations from their mean μ.

$$\therefore \quad \sigma^2 = \frac{\sum x^2}{n} - \mu^2 \text{ or } \frac{\sum fx^2}{\sum f} - \mu^2.$$

9 **Standard deviation** – the positive square root of the variance.

Test yourself

What to review

If your answer is incorrect:

1 The number of cars serviced by a small garage over a 12-week period are:

13 9 13 10 11 12 14 8 15 13 10 14

Find:

(a) the mode

(b) the median

(c) the lower quartile

(d) the 40th percentile

(e) the mean of the number of cars serviced per week.

Review Edexcel Book S1 pages 12–14 and 31–37

2 Each day a quality assurance manager records the number of defective items coming off one of the production lines for which he is responsible. The results for a period of 100 days are shown below.

Number of defective items	7	8	9	10	11	12	13	14	15
Number of days	5	12	15	24	17	13	10	3	1

(a) Write down the modal number of defective items produced in a day.

(b) Find the three quartiles for these data.

(c) Write down the 64th percentile.

(d) Calculate the mean number of defective items coming from this production line per day.

Review Edexcel Book S1 pages 12–14 and 31–37

3 Summarised in the table below are the lifetimes, to the nearest hour, of a random sample of 250 Superglow lightbulbs.

Lifetime	350–359	360–369	370–374	375–379
Number of bulbs	3	21	38	62

Lifetime	380–389	390–399	400–429
Number of bulbs	75	37	14

Calculate estimates of:

(a) the three quartiles **(b)** the 95th percentile

(c) the mean of these data.

Review Edexcel Book S1 pages 31–37

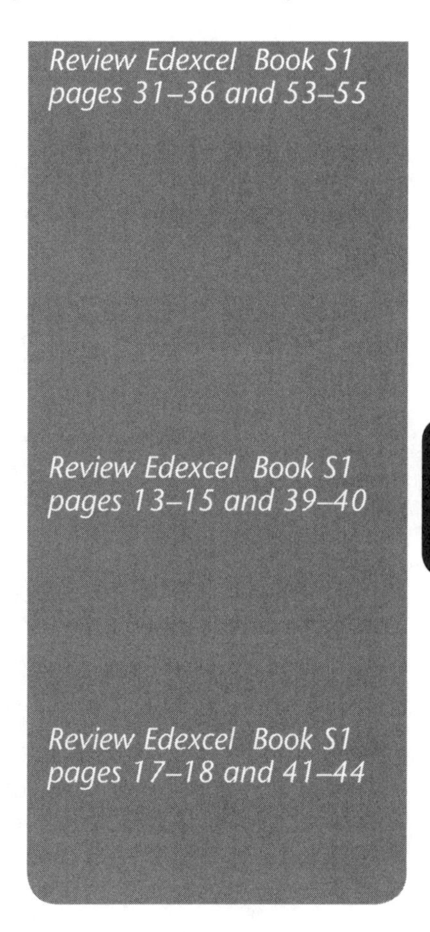

4 The number of scripts marked each day by an examiner are summarised in the following stem and leaf diagram.

Number of scripts 1|7 means 17

Stem	Leaf						
1	7	8					(2)
2	1	2	2	3	5	9	(6)
3	0	2	4	6	8	9	(6)
4	2	5	7				(3)

(a) Find the range of the data.

(b) Find the interquartile range.

5 The 200 members of a golf club each recorded the number, x, of rounds of golf played in a week. The results are summarised as follows:

$$\sum x = 527, \quad \sum x^2 = 1719$$

Find the mean and the standard deviation of the number of rounds of golf played by the members that week.

6 The following table shows the number of take-away meals eaten by a random sample of students during a particular month.

Number of meals	8	9	10	11	12	13	14
Number of students	2	5	12	8	5	3	2

Calculate the mean and the standard deviation of the numbers of take-away meals eaten by this sample.

Example 1

The following distribution shows the number of cars serviced by a garage during each week over a period of 40 weeks. Find:

(a) the mode

(b) the three quartiles

(c) the 37th percentile

(d) the mean.

Number of services	12	13	14	15	16	17	18	19	20
Number of weeks	2	4	4	6	10	5	4	3	2

This type of question is more easily answered using a table as follows:

No. of services (x)	No. of weeks (f)	Cumulative freq.	$f \times x$
12	2	2	24
13	4	6	52
14	4	10	56
15	6	16	90
16	10	26	160
17	5	31	85
18	4	35	72
19	3	38	57
20	2	40	40
	40		636

(a) Using **8** from Chapter 2, Mode = 16

Check that your answer is a value of the variable and not the modal frequency.

(b) Using **2**, since $n = 40$ then

for Q_1: $\frac{1}{4}n = 10 \Rightarrow$ use 10th and 11th observations

$\therefore \quad Q_1 = \frac{1}{2}(14 + 15) = 14.5$

For Q_2 (Median):

$\frac{1}{2}(n + 1) = 20.5 \Rightarrow$ use 20th and 21st observations

$\therefore \quad Q_2 = \frac{1}{2}(16 + 16) = 16$

For Q_3: $\frac{3}{4}n = 30 \Rightarrow$ Use 30th and 31st observations

$\therefore \quad Q_3 = \frac{1}{2}(17 + 17) = 17$

(c) Using **5**, for P_{37}:

$\frac{37}{100}n = 14.8 \Rightarrow$ use 15th observation.

$\therefore \quad P_{37} = 15$

(d) Using **10** from Chapter 2, Mean $= \bar{x} = \frac{636}{40} = 15.9$

Example 2

All 10 members of a jazz band each recorded, to the nearest mile, the number of miles travelled in a particular week to play in that week's sessions. The results are shown below.

20 17 9 32 12 15 25 10 16 22

(a) Find the mean number of miles travelled by the members that week.

(b) Explain why, in this situation, the mean is denoted by μ and not $\bar{x}$.

(c) Find the standard deviation, σ, of these data.

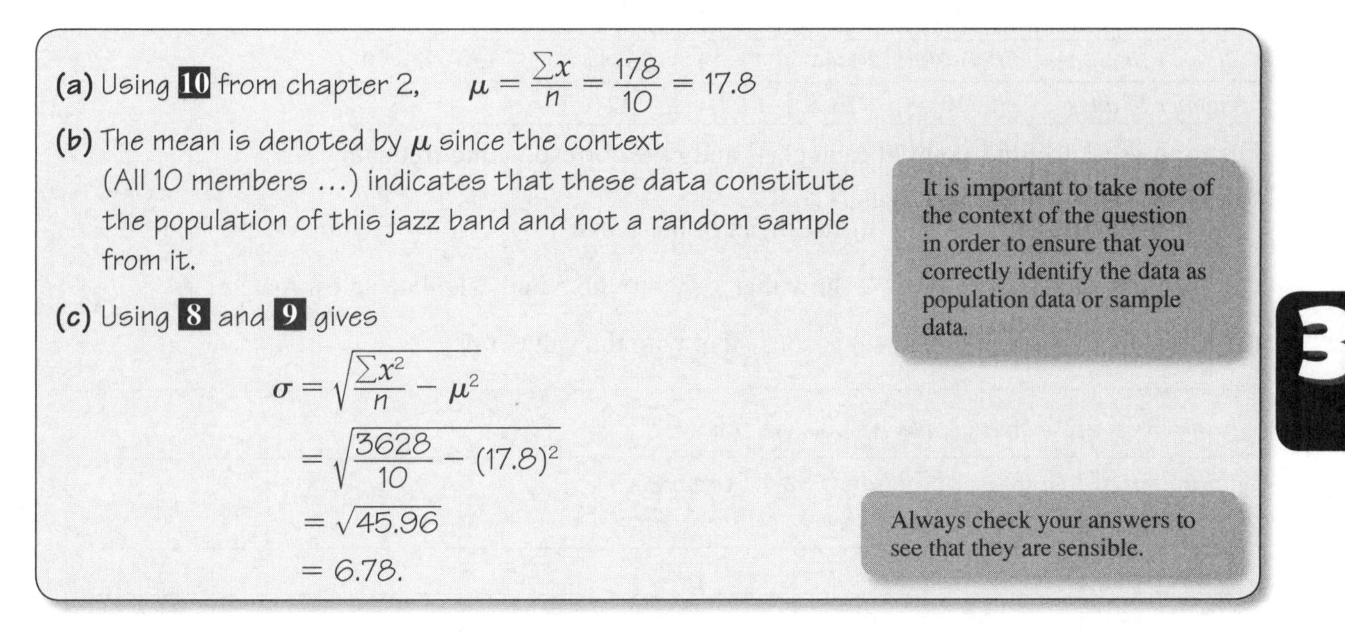

(a) Using **10** from chapter 2, $\mu = \frac{\sum x}{n} = \frac{178}{10} = 17.8$

(b) The mean is denoted by μ since the context (All 10 members ...) indicates that these data constitute the population of this jazz band and not a random sample from it.

It is important to take note of the context of the question in order to ensure that you correctly identify the data as population data or sample data.

(c) Using **8** and **9** gives

$$\sigma = \sqrt{\frac{\sum x^2}{n} - \mu^2}$$
$$= \sqrt{\frac{3628}{10} - (17.8)^2}$$
$$= \sqrt{45.96}$$
$$= 6.78.$$

Always check your answers to see that they are sensible.

Example 3

A market researcher at a garden centre asked a random sample of visitors to the garden centre how many hours, to the nearest hour, they had spent gardening last week. The results are summarised in the table below.

Hours of gardening	2–4	5–7	8	9	10–12	13–15	16–20
Number of visitors	3	7	10	12	5	4	1

(a) Write down an estimate of the range for these data, and explain why it is only an estimate.

(b) Find an estimate of the variance of these data.

(a) Using **1**, Range $= 20.5 - 1.5 = 19$
It is only an estimate because of the loss of accuracy when data is tabulated in a grouped frequency table. For example it is not possible to know what the 3 values in the first class are. To obtain an estimate the class boundaries are used.

(b) Let x represent the mid-point of the groups and f the frequency. Using mid-points 3, 6, 8, 9, 11, 14 and 18 gives $\sum f = 42$, $\sum fx = 368$ and $\sum fx^2 = 3604$.

Using **8**, $\sigma^2 = \frac{1}{\sum f}\left\{\sum fx^2 - \frac{(\sum fx^2)}{\sum f}\right\}$

$$= \frac{1}{42} - \left\{3604 - \frac{368^2}{42}\right\}$$
$$= 9.04$$

Example 4

The number of minutes, to the nearest minute, a family spent 'washing up' on a random sample of 100 days are summarised in the following table.

Number of minutes	30–39	40–44	45–49	50–54	55–59	60–79
Number of days	10	13	24	42	8	3

Let x represent the mid-point of each class and f the corresponding frequency.

(a) Calculate estimates of:
  (i) the three quartiles **(ii)** the mean of these data.

(b) Using the coding $y = \frac{x - 52}{5}$ show that $\sum fy = -66.5$ and calculate an estimate of $\bar{y}$.

(c) Show that the mean in part **(a)** is consistent with the value of $\bar{y}$.

Again it is advisable to use a table as follows:

Number of minutes	Number of days (f)	Mid-point (x)	Cumulative frequency	fx	$y = \frac{x-52}{5}$	fy
30–39	10	34.5	10	345	−3.5	−35
40–44	13	42	23	546	−2	−26
45–49	24	47	47	1128	−1	−24
50–54	42	52	89	2184	0	0
55–59	8	57	97	456	1	8
60–79	3	69.5	100	208.5	3.5	10.5
Totals	100			4867.5		−85 + 18.5 −66.5

If you do not use tables but try to use your calculator then remember that only correct answers gain marks. It is better to show your working.

Using **2**, **(a) (i)** $Q_1 = 44.5 + \frac{25 - 23}{24} \times 5 = 44.9$

$Q_2 = 49.5 + \frac{50 - 47}{42} \times 5 = 49.9$

$Q_3 = 49.5 + \frac{75 - 47}{42} \times 5 = 52.8$

Note that $\frac{n}{4}$ has been used in **(a) (i)**, although some text books use $\frac{(n+1)}{4}$. Both are acceptable in S1.

Always remember to use class boundaries

Using **10** from chapter 2, **(ii)** $\bar{x} = \frac{\sum fx}{\sum f} = \frac{4867.5}{100} = 48.675$

Using **11** from chapter 2, **(b)** $\bar{y} = \frac{-66.5}{100} = -0.665$

Note that $\sum fy$ is shown to be equal to −66.5 in the table.

Using **11** from chapter 2, **(c)** $\bar{x} = 5\bar{y} + 52 = 5 \times (-0.665) + 52 = 48.675$

This value is consistent with that found in **(a) (ii)**.

Worked examination question 1 [E]

A teacher monitored the time spent on homework by a random sample of 120 of her students. The times, to the nearest minute, are summarised in the following table.

Time	10–14	15–19	20–24	25–29	30–34
Number	1	4	16	30	28

Time	35–44	45–59	60–89	90–119
Number	27	8	4	2

By calculation, obtain an estimate of the median and quartiles of this distribution.

Time	Number (f)	Mid-point x	Cumulative frequency	$y = \frac{x - 27}{5}$	fy
10–14	1	12	1	−3	−3
15–19	4	17	5	−2	−8
20–24	16	22	21	−1	−16
25–29	30	27	51	0	0
30–34	28	32	79	1	28
35–44	27	39.5	106	2.5	67.5
45–59	8	52	114	5	40
60–89	4	74.5	118	9.5	38
90–119	2	104.5	120	15.5	31
Totals	120				177.5

Using **2**, $Q_1 = 24.5 + \frac{30 - 21}{30} \times 5 = 26$

$$Q_2 = 29.5 + \frac{60 - 51}{28} \times 5 = 31.1$$

$$Q_3 = 34.5 + \frac{90 - 79}{27} \times 10 = 38.6$$

Worked examination question 2 [E]

The 30 members of the Darton town orchestra each recorded the individual amount of practice, x hours, that they did in the first week in June. The results are summarised as follows:

$$\sum x = 225, \quad \sum x^2 = 1755$$

The mean and standard deviation of the number of hours of practice undertaken by the members of the Darton orchestra in this week were μ and σ respectively.

(a) Find μ. **(b)** Find σ.

Two new people joined the orchestra and the number of hours of individual practice they did in the first week of June were $\mu - 2\sigma$ and $\mu + 2\sigma$.

(c) State, giving your reasons, whether the effect of including these two members was to increase, decrease or leave unchanged the mean and standard deviation.

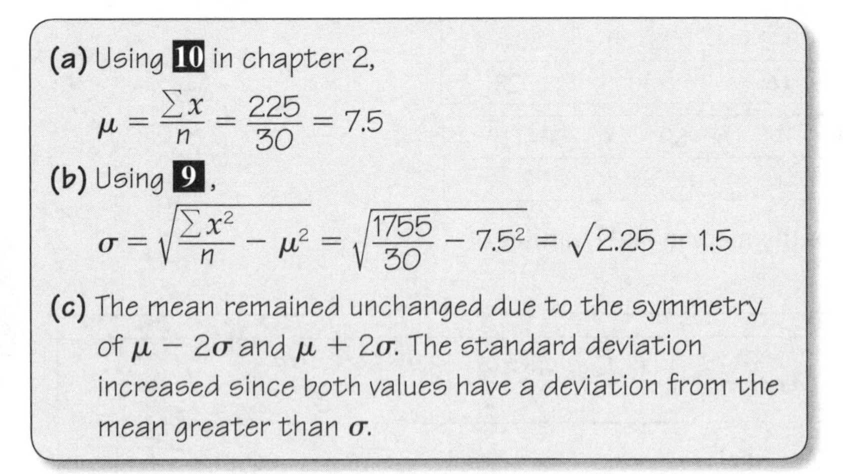

(a) Using **10** in chapter 2,

$$\mu = \frac{\sum x}{n} = \frac{225}{30} = 7.5$$

(b) Using **9**,

$$\sigma = \sqrt{\frac{\sum x^2}{n} - \mu^2} = \sqrt{\frac{1755}{30} - 7.5^2} = \sqrt{2.25} = 1.5$$

(c) The mean remained unchanged due to the symmetry of $\mu - 2\sigma$ and $\mu + 2\sigma$. The standard deviation increased since both values have a deviation from the mean greater than σ.

Revision exercise 3

1 The number of S1 books sold each week by a bookshop over a period of 15 weeks is as follows:

10 8 11 6 10 7 5 14 10 12 13 15 15 9 10

Find:

(a) the mode
(b) the median
(c) the upper quartile
(d) the 86th percentile
(e) the mean number of books sold per week.

2 The heights, to the nearest centimetre, of a random sample of adult males are summarised in the table below:

Height	*Number of adult males*
150–159	5
160–169	10
170–174	41
175–179	72
180–189	16
190–199	6

Calculate estimates of:

(a) the three quartiles
(b) the 90th percentile
(c) the mean of these data.

3 For the data in question **2**, let x represent the mid-point of each class and f the corresponding frequency.

(a) Show that using the coding $y = \frac{x - 177}{5}$, $\sum fy = -43.5$

(b) Find $\bar{y}$.

(c) Use your value of $\bar{y}$ to find $\bar{x}$ and show that this gives the same value as you obtained in part **(c)** of question **2**.

4 The salesmen at a garage recorded the number of cars each of them sold during a particular month. The results are shown below:

Hint: think about whether this is population or sample data.

17 21 19 15 26 22 18 24

Calculate: **(a)** the mean **(b)** the standard deviation of the number of cars sold that month.

3

5 A random sample of 30 students each recorded the number, x, of videos they watched during a three-week period. The results are summarised below:

$$\sum x = 407, \qquad \sum x^2 = 6535$$

Calculate the standard deviation of these data.

Representation of data 4

What you should know

1 **Stem and leaf diagram** – each row represents a stem and is indicated by the number to the left of the vertical line. The digits to the right of the vertical line are leaves associated with the stem.

5 | 0 means 50

stem		leaves
5	0 1 1 2 4	
6	0 1 2 6	
7	1 2 4 8	

2 **Outlier** – an extreme value which does not fit into the main body of the data. Any rules to identify outliers will be specified in the S1 question paper.

3 **Histogram** – used to represent a continuous variable which has been summarised by a group frequency distribution. Each group or class is represented by a bar with width equal to the class width and area proportional to its frequency.

The variable goes along the horizontal axis and the vertical axis is frequency density.

Class	*Frequency*	*Class width*	*Frequency density* $= \dfrac{\textit{Frequency}}{\textit{class width}}$
7.5 8–10 10.5	12	3	4

Relative frequency and relation frequency histograms will not be examined in S1.

The histogram is plotted using class boundaries and frequency densities.

4 **Relative frequency** – a frequency expressed as a proportion of the total frequency.

5 **Relative frequency histogram** – a histogram based on relative frequencies.

6 **Box plot** – used to represent data and allows comparisons.

Smallest value Q_1 Q_2 Q_3 Largest value

Questions involving the drawing of simple box plots will not be set in S1.

7 **Skewness** – an indicator of the shape of a distribution.

Symmetrical: mode = median = mean

or $Q_2 - Q_1 = Q_3 - Q_2$

Positive skew: mode < median < mean

or $Q_2 - Q_1 < Q_3 - Q_2$

Negative skew: mean < median < mode

or $Q_2 - Q_1 > Q_3 - Q_2$

Test yourself

What to review

If your answer is incorrect:

1 A city operates a 'Park and Ride' bus service from a particular car park into the city and back. As part of an exercise to assess the success of the service the manager recorded the number of passengers getting off each of the buses arriving back at the car park during one day. The following are the results.

10	16	35	29	30	32	37	20	9	41
17	11	23	19	21	34	47	11	27	40
42	24	10	17	49	23	6	35	28	37

(a) Construct a stem and leaf diagram to summarise these data.

(b) Suggest why these data will be of limited use to the manager in assessing the success of the service.

Review Edexcel Book S1 pages 53–56

2 A couple shop regularly at their local supermarket. Over a period of weeks they recorded, to the nearest minute, the number of minutes spent shopping on each occasion they visited it. Their data are summarised in the following grouped frequency distribution.

Number of minutes	20–29	30–39	40–44	45–49	50–59	60–79
Number of visits	6	9	12	17	5	2

(a) For the first class write down the class boundaries, mid-point and class width.

(b) Calculate the frequency densities needed to plot a histogram of these data.

Review Edexcel Book S1 pages 9–12 and 61–66

3 The table gives the ages, in complete years, of the population in a particular region of the United Kingdom.

Age	0–4	5–15	16–44	45–64	65–79	80 and over
Number (in thousands)	260	543	1727	756	577	135

A histogram of these data was drawn with age along the horizontal axis. The 0–4 age group was represented by a bar of horizontal width 0.5 cm and height 5.2 cm.

Find, in cm to 1 decimal place, the widths and heights of the bars representing the following age groups:

(a) 16–44 **(b)** 65–79.

Review Edexcel Book S1 pages 61–66

4 The following stem and leaf diagram summarises the number of CDs sold by a music shop over a 50-day period.

Review Edexcel Book S1 pages 12–14, 31–37, 58–59

Number of CDs   5|0 means 50

Stem	Leaves	
5	0 1 1 2	(4)
5	5 5 6 7 8 9	(6)
6	0 0 1 2 3 4 4	(7)
6	5 7 7 7 7 8 9 9	(8)
7	0 0 1 2 4 4	(6)
7	5 5 6 6 7 9	(6)
8	0 1 1 2 2 4	(6)
8	5 5 6 8	(4)
9	0 1 3	(3)

(a) Write down the modal number of CDs sold.

(b) Find the three quartiles for these data.

(c) Draw a box plot to represent these data.

(d) Find the mean number of CDs sold by the shop during this 50-day period.

5 Data were collected such that $Q_1 = 26$, $Q_2 = 30$ and $Q_3 = 38$.

Review Edexcel Book S1 pages 58–59

(a) Using the $1.5(Q_3 - Q_1)$ rule, find the values outside which data was classified as being outliers.

The values less than Q_1 were 17, 18, and 21, and those greater than Q_3 were 52, 60 and 64.

(b) Draw a box plot to represent these data.

6 Whig and Penn, solicitors, monitored the time spent on consultations with a random sample of 120 of their clients. The times, to the nearest minute, are summarised in the following table.

Review Edexcel Book S1 pages 20–22, 41–44 and 60–68

Time	10–14	15–19	20–24	25–29	30–34
Number of clients	2	5	17	33	27

Time	35–44	45–59	60–89	90–119
Number of clients	25	7	3	1

(a) By calculation, obtain estimates of the median and quartiles of this distribution.

(b) Comment on the skewness of the distribution.

(c) Explain briefly why these data are consistent with the distribution of times you might expect in this situation.

(d) Calculate the mean and variance of these data.

The solicitors are undecided whether to use the median and quartiles, or the mean and standard deviation to summarise these data.

(e) State, giving a reason, which you would recommend them to use.

(f) Given that the least time spent with a client was 12 minutes and the longest time was 116 minutes, draw a box plot to represent these data. Use graph paper and show your scale clearly.

Law and Court, another group of solicitors, monitored the times spent with a random sample of their clients. They found that the least time spent with a client was 20 minutes, the longest time was 40 minutes and the quartiles were 24, 30 and 36 minutes respectively.

(g) Using the same graph paper and the same scale draw a box plot to represent these data.

(h) Compare and contrast the two box plots. [E]

Example 1

A leisure club manager recorded the number of people using the club each day over a long period of time. The results are summarised as follows.

Smallest number = 21, lower quartile = 36, median = 102, upper quartile = 130, largest number = 152.

(a) Find the range of these data.

(b) Find the interquartile range.

(c) Comment on the skewness of these data.

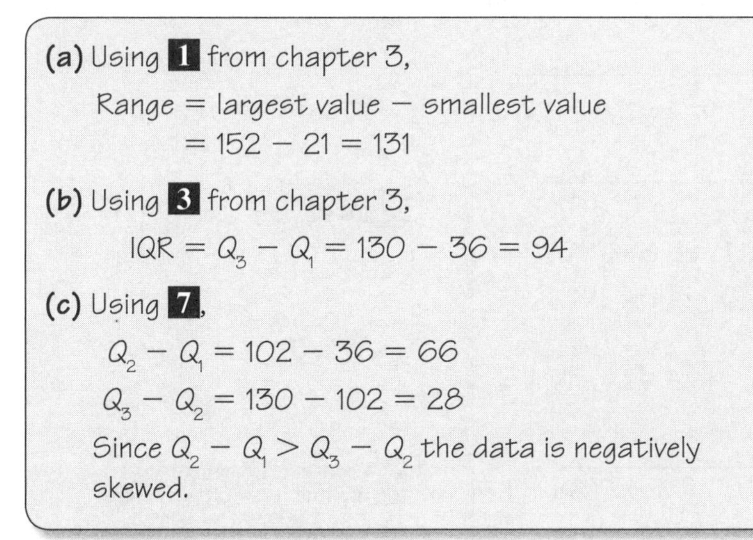

(a) Using **1** from chapter 3,

Range = largest value − smallest value
= 152 − 21 = 131

(b) Using **3** from chapter 3,

$IQR = Q_3 - Q_1 = 130 - 36 = 94$

(c) Using **7**,

$Q_2 - Q_1 = 102 - 36 = 66$

$Q_3 - Q_2 = 130 - 102 = 28$

Since $Q_2 - Q_1 > Q_3 - Q_2$ the data is negatively skewed.

Note that in statistics the range is a number whereas in pure maths it might be expressed as $y \in \mathbb{R}$ or $-7 \leq y \leq 11$ for example.

Example 2

Data were collected by a student such that

$$Q_1 = 27, Q_2 = 32 \text{ and } Q_3 = 39.$$

(a) Using $1.5(Q_3 - Q_1)$ find the values beyond which any outliers in the data will lie.

The data included the three outliers: 5, 60, and 65.

(b) Draw a box plot to represent the student's data.

(a) Using **2**, $1.5(Q_3 - Q_1) = 1.5(39 - 27) = 18$

$\therefore$ Values are $27 - 18 = 9$ and $39 + 18 = 57$.

(b) Using **6**,

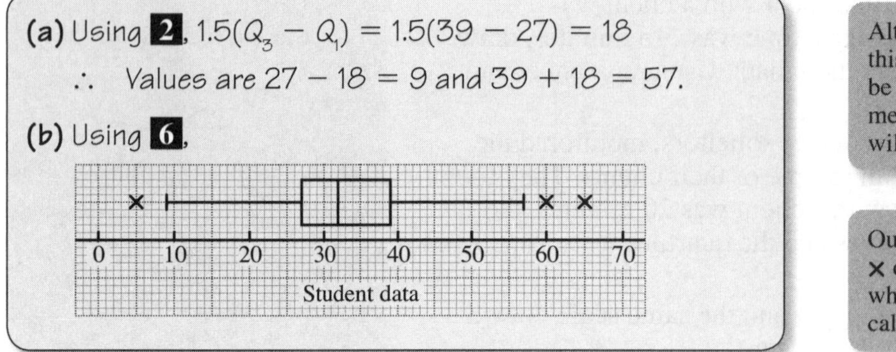

Although $1.5(Q_3 - Q_1)$ is used in this example this may not always be the case. On S1 papers the method of identifying outliers will be defined.

Outliers are usually shown by × outside the 'whiskers'. The whiskers show the extremes calculated in part **(a)**.

Example 3

An ornithologist recorded the number of species of bird seen on each of 40 bird watching trips.

37	13	18	45	25	41	32	15	42	38
36	19	21	41	34	37	41	29	21	32
18	35	42	22	36	19	25	38	41	26
35	22	40	20	25	31	21	24	20	33

(a) Construct a stem and leaf diagram for these data.

(b) Give a disadvantage associated with constructing stem and leaf diagrams.

(a) Unordered

Species 1|8 means 18

1	8 3 9 8 9 5	(6)
2	2 1 2 0 5 5 5 1 9 4 1 0 6	(13)
3	7 6 5 5 4 6 7 1 2 8 8 2 3	(13)
4	2 0 5 1 1 1 2 1	(8)

Using **1**

Ordered

Species 1|3 means 13

1	3 5 8 8 9 9	(6)
2	0 0 1 1 1 2 2 4 5 5 5 6 9	(13)
3	1 2 2 3 4 5 5 6 6 7 7 8 8	(13)
4	0 1 1 1 1 2 2 5	(8)

Note that to avoid very long stems they may be split.

(b) Construction time to create an ordered diagram can be lengthy.

Example 4

A random sample of 45 students sat a chemistry examination. The marks they gained out of 100 are represented on the box plot below.

45 55 65 75 85

Mark

(a) Write down:
 (i) the lowest mark, (ii) the highest mark,
 (iii) the mark exceeded by 25% of the students.

(b) Explain briefly why box plots can be useful when several similar sets of data have been collected.

(a) (i) 50 (ii) 81 (iii) 77.

(b) Allows for easy comparisons provided all the box plots use the same scale.

Example 5

A random sample of 50 students in a college recorded the number of children in their family. The table below summarises the results.

Number of children (x)	1	2	3	4	5	6
Number of students (f)	8	17	11	10	2	2

(a) Write down the range for the above distribution.

(b) Find the interquartile range of these data.

(c) Calculate the standard deviation for this set of numbers.

(d) Comment on the skewness of this distribution.

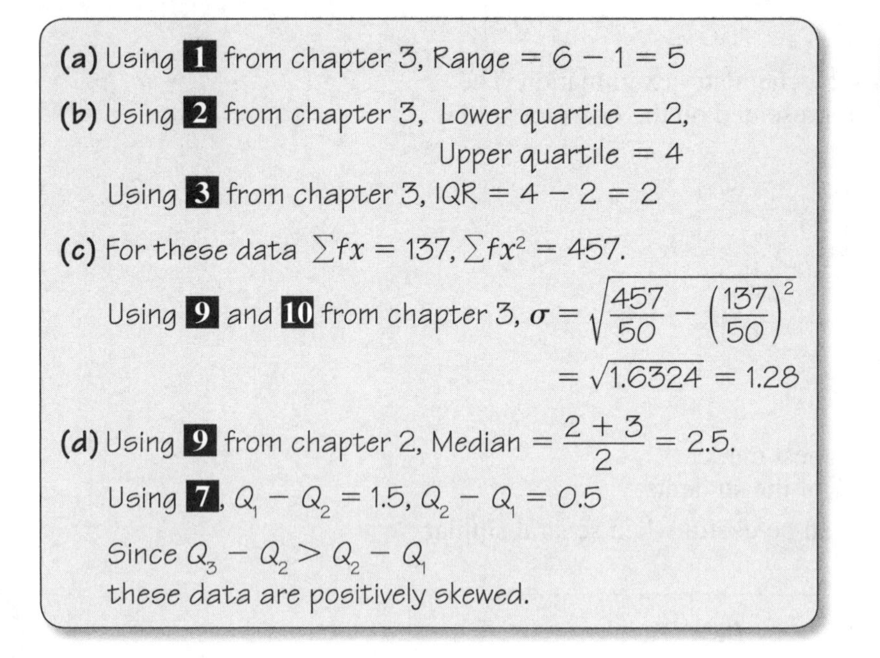

(a) Using **1** from chapter 3, Range = 6 − 1 = 5

(b) Using **2** from chapter 3, Lower quartile = 2,
Upper quartile = 4

Using **3** from chapter 3, IQR = 4 − 2 = 2

(c) For these data $\sum fx = 137, \sum fx^2 = 457$.

Using **9** and **10** from chapter 3, $\sigma = \sqrt{\frac{457}{50} - \left(\frac{137}{50}\right)^2}$
$= \sqrt{1.6324} = 1.28$

(d) Using **9** from chapter 2, Median $= \frac{2+3}{2} = 2.5$.

Using **7**, $Q_1 - Q_2 = 1.5$, $Q_2 - Q_1 = 0.5$

Since $Q_3 - Q_2 > Q_2 - Q_1$
these data are positively skewed.

Note that the range is found using the values of *x*.

Worked examination question 1 [E]

The children in classes *A* and *B* were each given a set of arithmetic problems to solve. Their times, to the nearest minute, were recorded and they are summarised in the box plot opposite.

Compare and contrast the results for the two classes.

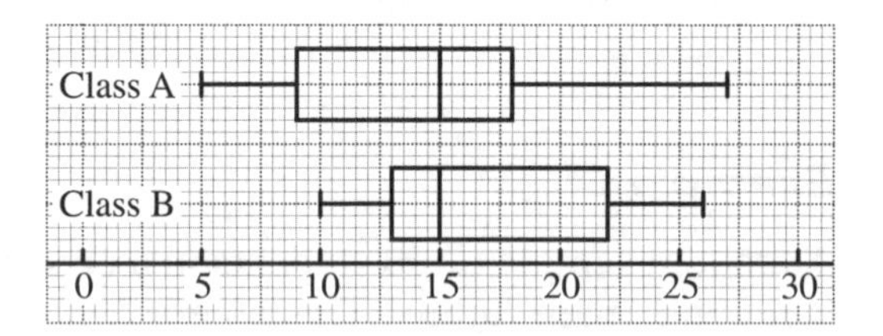

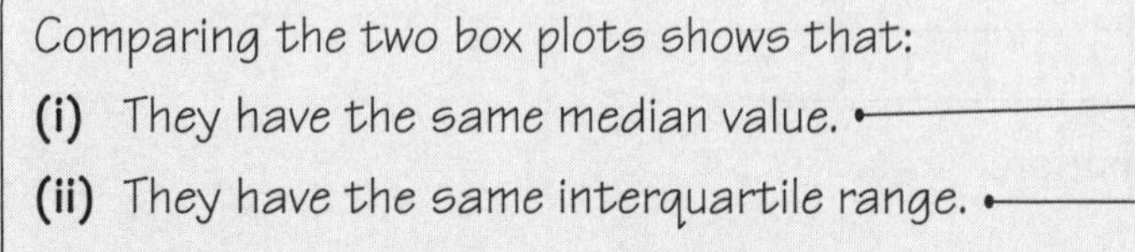

Comparing the two box plots shows that:

(i) They have the same median value.

(ii) They have the same interquartile range.

(iii) *A* is negatively skewed whereas *B* is positively skewed.

(iv) The range of *B* is smaller than the range of *A*.

Using **9** in chapter 2

Using **3** in chapter 3

Using **7**

Using **1** in chapter 3

Worked examination question 2 [E]

The table summarises the birth weights of a random sample of 100 babies.

Birth weight (kg)	1.0–	1.4–	1.6–	1.8–	2.0–
Number of babies	12	17	23	14	10

Birth weight (kg)	2.2–	2.4–	2.8–	3.2–3.8
Number of babies	9	8	4	3

(a) Write down the upper class boundary of the first class.

(b) Explain why it is appropriate to represent these data by a bar chart.

(c) State the principle involved when drawing a histogram.

(d) Represent these data by a histogram.

(e) Estimate the number of babies with birth weight, recorded to 2 s.f., between 2.1 and 2.5 kg.

(a) The upper boundary is 1.4

Using **7** from chapter 2

(b) Weight is a continuous variable.

(c) For each bar, and hence the whole histogram, area is proportional to frequency.

(d)

Birth weight	Number of babies	Class width	Frequency density
1.0–	12	0.4	30
1.4–	17	0.2	85
1.6–	23	0.2	115
1.8–	14	0.2	70
2.0–	10	0.2	50
2.2–	9	0.2	45
2.4–	8	0.4	20
2.8–	4	0.4	10
3.2–3.8	3	0.6	5

Data summarised like this (1.4–, etc) means that the values shown are class boundaries.

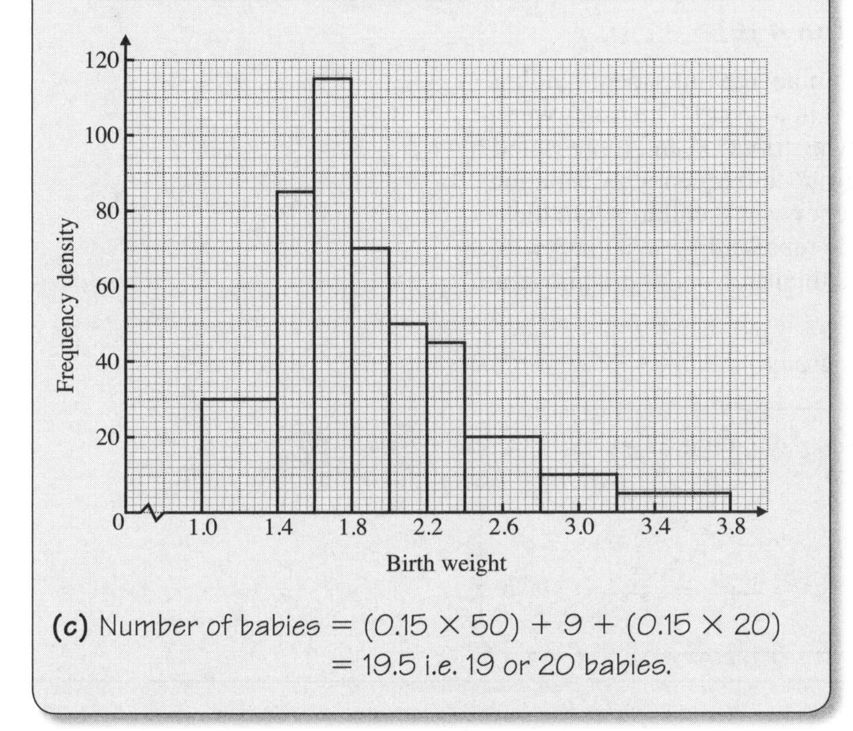

Using **3**

(c) Number of babies $= (0.15 \times 50) + 9 + (0.15 \times 20)$
$= 19.5$ i.e. 19 or 20 babies.

0.15 is used since 2.1 kg is assumed to mean 2.05–2.15

Worked examination question 3 [E]

The scores on a psychological test obtained by a random sample of 50 job applicants are shown below.

39	53	63	75	90	35	51	63	72	85
25	44	57	68	78	31	48	60	71	82
34	49	61	72	83	28	46	59	69	80
22	44	56	68	78	39	53	63	74	88
42	55	66	76	96	40	55	65	75	92

(a) Construct a stem and leaf diagram to represent these data.

(b) Give an advantage of using such a diagram.

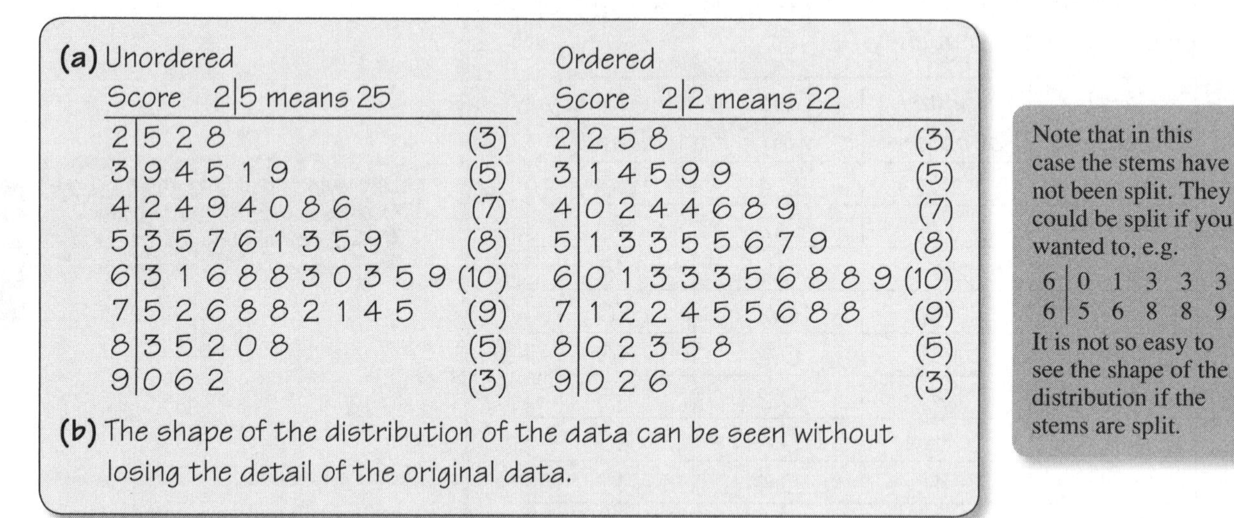

(a) Unordered

Score 2|5 means 25

Stem	Leaves	
2	5 2 8	(3)
3	9 4 5 1 9	(5)
4	2 4 9 4 0 8 6	(7)
5	3 5 7 6 1 3 5 9	(8)
6	3 1 6 8 8 3 0 3 5 9	(10)
7	5 2 6 8 8 2 1 4 5	(9)
8	3 5 2 0 8	(5)
9	0 6 2	(3)

Ordered

Score 2|2 means 22

Stem	Leaves	
2	2 5 8	(3)
3	1 4 5 9 9	(5)
4	0 2 4 4 6 8 9	(7)
5	1 3 3 5 5 6 7 9	(8)
6	0 1 3 3 3 5 6 8 8 9	(10)
7	1 2 2 4 5 5 6 8 8	(9)
8	0 2 3 5 8	(5)
9	0 2 6	(3)

(b) The shape of the distribution of the data can be seen without losing the detail of the original data.

Note that in this case the stems have not been split. They could be split if you wanted to, e.g.

6	0 1 3 3 3
6	5 6 8 8 9

It is not so easy to see the shape of the distribution if the stems are split.

Worked examination question 4 [E]

A teacher recorded, to the nearest minute, the time spent reading during a particular day by each child in a group. The times were summarised in a grouped frequency distribution and represented by a histogram. The first class in the grouped frequency distribution was 10–19 and its associated frequency was 8 children. On the histogram the height of the rectangle representing that class was 2.4 cm and the width was 2 cm. The total area under the histogram was 53.4 cm^2.

Find the number of children in the group.

8 children are represented by a rectangle of area $2 \times 2.4 = 4.8\,\text{cm}^2$.

$\therefore$ 1 child is represented by $\frac{4.8}{8} = 0.6\,\text{cm}^2$.

$\therefore$ Total number of children $= \dfrac{\text{Total area}}{0.6} = \dfrac{53.4}{0.6} = 89$

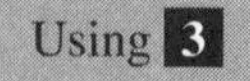

Worked examination question 5 [E]

The following table summarises the results of a sales manager's analysis of the amounts to the nearest £, of a sample of 750 invoices.

Amount (£)	0–9	10–19	20–49	50–99	100–149	150–199	200–499	500–749
Number of invoices	50	204	165	139	75	62	46	9

Let x represent the mid-point of each class.

Thus with $y = \frac{x - 14.5}{10}$, $\sum fy = 5021$ and $\sum fy^2 = 115\,773.5$.

(a) Using these values, or otherwise, find estimates of the mean and the standard deviation of these invoices.

(b) Explain why the mean and the standard deviation might not be the best summary statistics to use with these data.

(c) Calculate estimates of alternative summary statistics which might be used by the sales manager.

(d) Use these estimates to justify your explanation in part **(b)**.

4

(a) Using **10** in chapter 2, $\bar{y} = \frac{5021}{750}$

Using **11** in chapter 2, $x = 10y + 14.5 \Rightarrow \bar{x} = £81.45$

Using **8** in chapter 3, $\sigma_y^2 = \frac{1}{750}\left\{115\,773.5 - \frac{5021^2}{750}\right\} = 109.546\,10$

$\therefore \qquad \sigma_x = 10\sigma_y = 10 \times \sqrt{109.546\,10} = £104.66$

(b) Looking at the data in the table and using **7** you can see that the data is positively skewed. In addition there are some large values on the invoices and these will tend to inflate the values of the mean and standard deviation.

(c) Alternative summary statistics are the median (Q_2) and the interquartile range ($Q_3 - Q_1$).

Using **9** in chapter 2 and **2** in chapter 3,

$$Q_1 = 9.5 + \frac{(187.5 - 50)}{204} \times 10 = 16.2$$

$$Q_2 = 19.5 + \frac{(375 - 254)}{165} \times 30 = 41.5$$

$$Q_3 = 99.5 + \frac{(562.5 - 558)}{75} \times 50 = 102.5$$

Thus the median = 41.5 and the interquartile range = 86.3

(d) The median is less than the mean and is not influenced by any of the extreme values and the interquartile range uses only 50% of the observations and is also not influenced by any of the extreme values. It is advisable to use the median and IQR with skew data and for these data $Q_3 - Q_2 > Q_2 - Q_1$ (i.e. $61 > 25.3$) confirming that the data is positively skewed.

Revision exercise 4

1 The following data relates to the number of daisies found in each of 15 randomly chosen 1 m^2 plots in a large field.

25 17 41 26 32 21 57 36 29 64 37 29 30 26 12

(a) Using the $1.5(Q_3 - Q_1)$ rule, find the values beyond which any outliers in these data lie.

(b) Identify any outliers.

2 The journey times to and from college for 30 students are shown in the two box plots below.

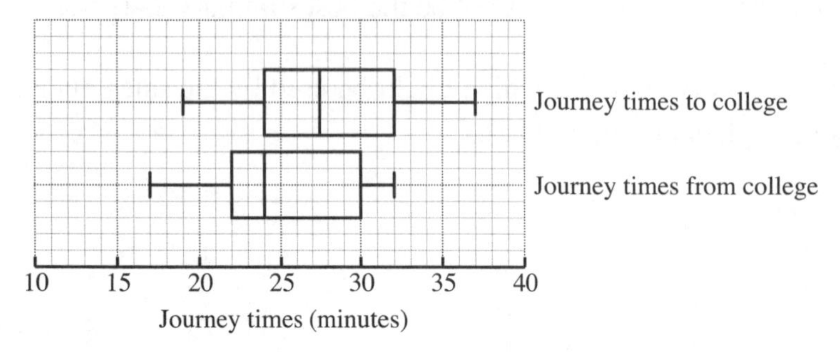

(a) Are journey times from college generally more or less than journey times to college?

(b) What was, **(i)** the shortest journey time, **(ii)** the longest journey time?

3 Hospital records show the number of babies born in a year. The number of babies delivered by 15 male doctors is summarised by the stem and leaf diagram below.

Babies		Totals	Babies		Totals
1	9	(1)	5	1	(1)
2	1 6 7 7	(4)	6	0	(1)
3	2 2 3 4 8	(5)	7		(0)
4	5	(1)	8	6 7	(2)

(a) Find the median and inter-quartile range of these data.

(b) Given that there are no outliers, draw a box plot on graph paper to represent these data. Start your scale at the origin.

(c) Calculate the mean and standard deviation of these data.

The records also contain the number of babies delivered by 10 female doctors.

34 30 20 15 6 32 26 19 11 4

The quartiles are 11, 19.5 and 30.

(d) Using the same scale as in part **(b)** and on the same graph paper draw a box plot for the data for the 10 female doctors.

(e) Compare and contrast the box plots for the data for male and female doctors. [E]

4 As part of her project a student selected a random sample of 40 students and each student was asked how many CDs they owned. The results were as shown below.

52	89	75	66	43	83	51	75	86	71
58	66	45	73	55	68	88	65	61	76
43	75	47	78	56	79	76	31	83	61
52	72	62	68	88	34	67	51	64	52

(a) Construct a stem and leaf diagram to represent these data.

(b) Write down how many students had fewer than 50 CDs.

(c) Comment on the distribution of CDs among this sample.

5 The following table summarises the weight, to the nearest kg, of a random sample of students.

Weight (kg)	45–54	55–59	60–64	65–69	70–74	75–79	80–89	90–104
Number of students	6	9	20	14	10	7	5	3

(a) Explain why a histogram is suitable to represent these data.

(b) Draw the histogram.

(c) Estimate the number of students weighing between 61 kg and 71 kg.

Probability 5

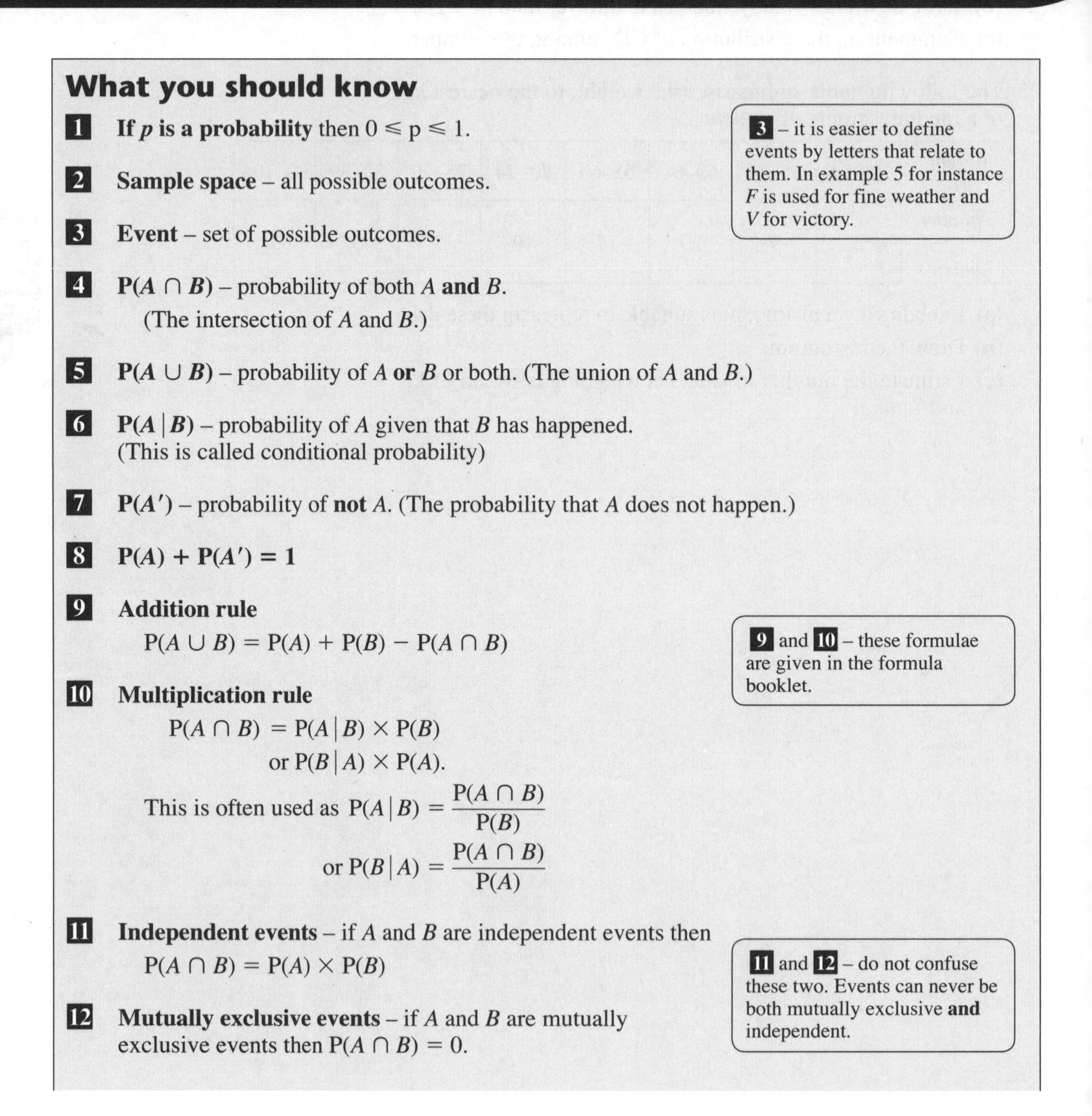

What you should know

1 **If p is a probability** then $0 \leqslant p \leqslant 1$.

2 **Sample space** – all possible outcomes.

3 **Event** – set of possible outcomes.

4 **$P(A \cap B)$** – probability of both A **and** B.
(The intersection of A and B.)

5 **$P(A \cup B)$** – probability of A **or** B or both. (The union of A and B.)

6 **$P(A \mid B)$** – probability of A given that B has happened.
(This is called conditional probability)

7 **$P(A')$** – probability of **not** A. (The probability that A does not happen.)

8 **$P(A) + P(A') = 1$**

9 **Addition rule**

$$P(A \cup B) = P(A) + P(B) - P(A \cap B)$$

10 **Multiplication rule**

$$P(A \cap B) = P(A \mid B) \times P(B)$$
$$\text{or } P(B \mid A) \times P(A).$$

This is often used as $P(A \mid B) = \dfrac{P(A \cap B)}{P(B)}$

or $P(B \mid A) = \dfrac{P(A \cap B)}{P(A)}$

11 **Independent events** – if A and B are independent events then
$P(A \cap B) = P(A) \times P(B)$

12 **Mutually exclusive events** – if A and B are mutually exclusive events then $P(A \cap B) = 0$.

3 – it is easier to define events by letters that relate to them. In example 5 for instance F is used for fine weather and V for victory.

9 and **10** – these formulae are given in the formula booklet.

11 and **12** – do not confuse these two. Events can never be both mutually exclusive **and** independent.

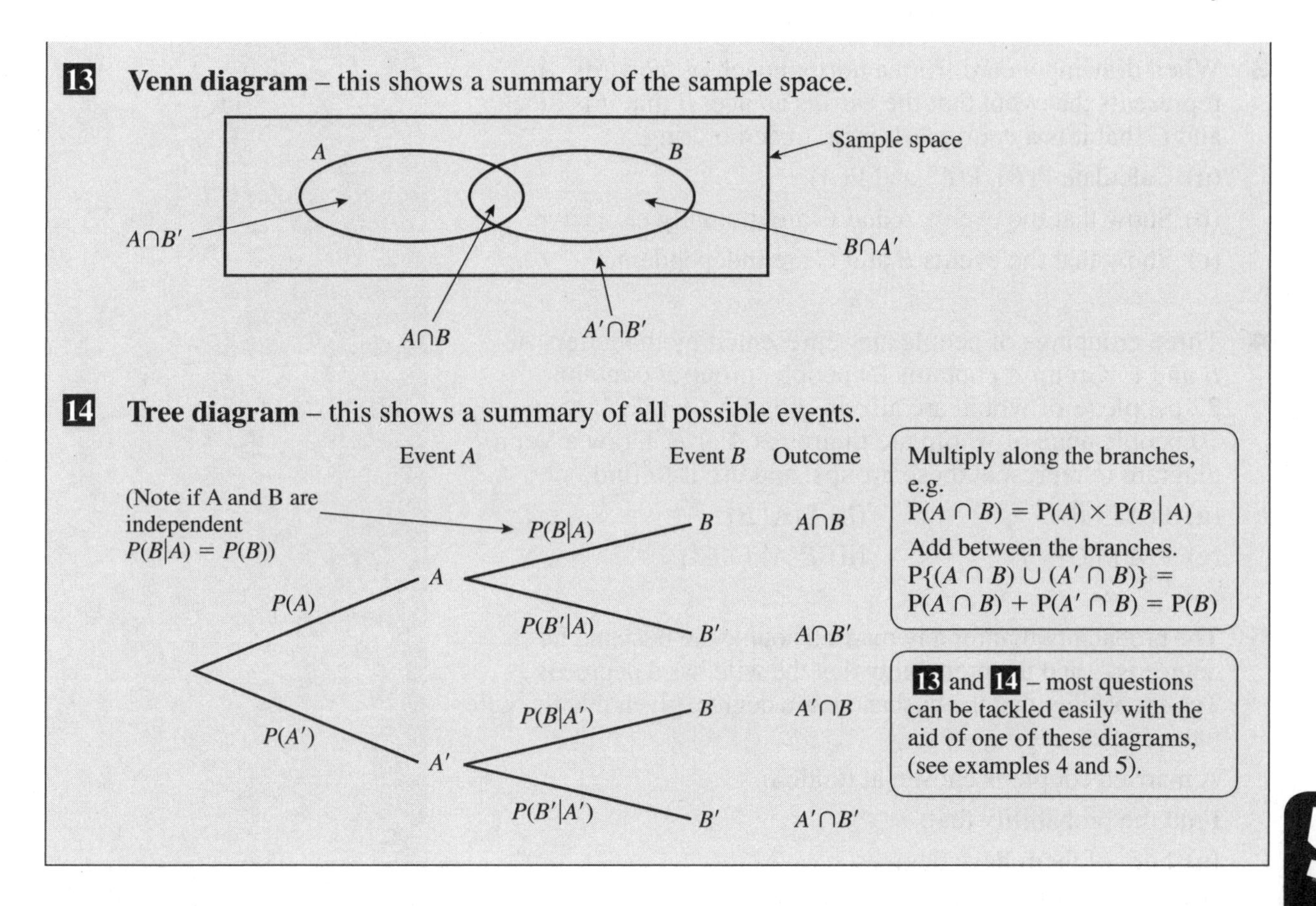

5

Test yourself

What to review

1 A bag contains 16 balls, numbered from 1 to 16. Some of the balls are yellow while the rest are red.

A ball is drawn at random from the bag.

(a) Write down the probability that a number greater than 12 will be drawn.

(b) Given that a prime number is a number that is only divisible by itself and 1, find the probability that the ball will be a prime number. (Note: the number 1 is not a prime number.)

(c) Given that the probability of selecting a red ball is $\frac{5}{16}$, find the probability of drawing a yellow ball.

If your answer is incorrect:
Review Edexcel Book S1 pages 78–79

2 The probability that a person is left handed is $\frac{2}{5}$ and the probability that a person wears glasses is $\frac{1}{4}$. Assuming that the two events are independent, find the probability that a person chosen at random is left-handed and wears glasses.

Review Edexcel Book S1 pages 95–100

3 When drawing a card from a normal pack of 52 cards, A represents the event that the card is an ace, B that it is black and C that it is a court card (jack, queen or king).

(a) Calculate $P(B)$, $P(C)$ and $P(A)$.

(b) Show that the events A and C are mutually exclusive.

(c) Show that the events B and C are independent.

Review Edexcel Book S1 pages 95–100

4 Three groupings of people are represented by the letters A, B and C. Group A contains 20 people, group B contains 22 people (8 of whom are also in group A), while C contains 10 people none of whom are in groups A or B. Draw a Venn diagram to represent these groups, and use it to find:

(a) $P(A \cap B)$ **(b)** $P(A \mid B)$

(c) $P(A \cap C)$ **(d)** $P(A' \cap C')$

Review Edexcel Book S1 pages 80–90

5 The probability that for any married couple the husband has a degree is $\frac{6}{10}$ and the probability that the wife has a degree is $\frac{1}{2}$. The probability that the husband has a degree, given that the wife has a degree, is $\frac{11}{12}$.

A married couple is chosen at random.

Find the probability that:

(a) both of them have degrees

(b) only one of them has a degree

(c) neither of them has a degree.

Two married couples are chosen at random.

(d) Find the probability that only one of the two husbands and only one of the two wives have a degree. [E]

Review Edexcel Book S1 pages 80–94

Example 1

Event A is the event of selecting an ace from a pack of playing cards. Event B is the event of selecting a black card. A pack of playing cards contains 52 cards – 13 each of diamonds, clubs, hearts and spades. Hearts and diamonds are red cards; spades and clubs are black cards.

Find P(A), P(B), P(B').

$P(A) = \frac{4}{52} = \frac{1}{13}$

$P(B) = \frac{26}{52} = \frac{1}{2}$

Using **8**, $P(B') = 1 - P(B) = 1 - \frac{1}{2} = \frac{1}{2}$

There are 4 aces in a pack of cards.

There are $2 \times 13 = 26$ black cards.

There are $2 \times 13 = 26$ red cards.

Example 2

The probability that event A occurs is 0.3 and the probability that event B occurs is 0.6. Find the probability of either or both events A and B happening given that:

(a) the events A and B are mutually exclusive

(b) the probability of both A and B happening is 0.2

(c) A and B are independent.

(a) If A and B are mutually exclusive $P(A \cap B) = 0$

$$\begin{aligned} P(A \cup B) &= P(A) + P(B) - P(A \cap B) \\ &= P(A) + P(B) - 0 \\ &= 0.3 + 0.6 \\ &= 0.9 \end{aligned}$$

(b) $P(A \cup B) = P(A) + P(B) - P(A \cap B)$

$= 0.3 + 0.6 - 0.2 = 0.7$

(c) If A and B are independent $P(A \cap B) = P(A) \times P(B)$

$$\begin{aligned} P(A \cup B) &= P(A) + P(B) - P(A \cap B) \\ &= P(A) + P(B) - P(A) \times P(B) \\ &= 0.3 + 0.6 - 0.3 \times 0.6 \\ &= 0.72 \end{aligned}$$

Example 3

It may be assumed that each birth in a family is equally likely to result in a boy or a girl.

(a) Comment on this assumption.

There are four children in a family.

(b) Given that at least one child is a boy, find the probability that there are exactly two boys.

(c) State another assumption made in part **(b)**.

(a) Let P(B) be the probability of a boy and P(G) be the probability of a girl.

$P(B) = P(G) = \frac{1}{2}$ is not necessarily realistic. Data collected by the national census would suggest that this is not so. It is, however, near enough to give a good model.

(b) Let A represent the event of at least one boy.
Let T represent the event of exactly two boys.

$$P(T|A) = \frac{P(A \cap T)}{P(A)}$$

$$= \frac{P(T)}{P(A)}$$

$$P(A) = 1 - P(A') = 1 - \left(\tfrac{1}{2}\right)^4 = \tfrac{15}{16}$$

$$P(T) = 6\left(\tfrac{1}{2}\right)^4 = \tfrac{3}{8}$$

$$P(T|A) = \frac{\frac{3}{8}}{\frac{15}{16}} = \frac{2}{5}$$

(c) Births are independent of one another.

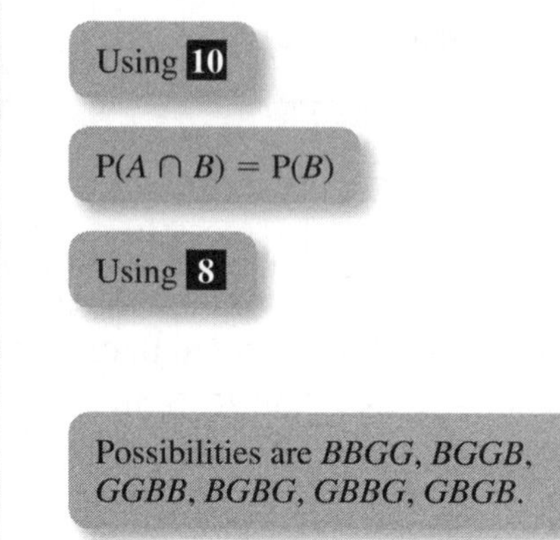

Example 4

Draw a Venn diagram for the events A and B as defined in example 1. Use it to find:

(a) $P(A|B)$ **(b)** $P(B'|A')$

Using **13**,

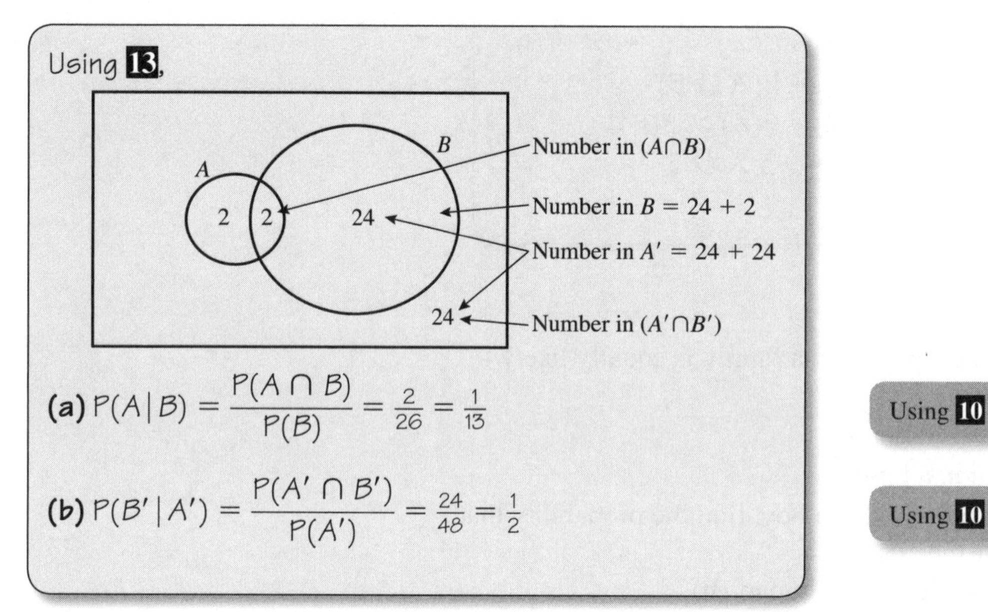

(a) $P(A|B) = \frac{P(A \cap B)}{P(B)} = \frac{2}{26} = \frac{1}{13}$

(b) $P(B'|A') = \frac{P(A' \cap B')}{P(A')} = \frac{24}{48} = \frac{1}{2}$

Using **10**

Using **10**

Example 5

The first-eleven hockey team is to play an important match. Past performance suggests that if the weather is fine the chance of victory is $\frac{4}{5}$ but if the weather is not fine then the chance of victory is only $\frac{1}{3}$. The forecasters say that the probability of fine weather for match day is $\frac{3}{4}$. Find the probability that the team has a victory.

Let F be the event of a fine day and V the event of a victory.

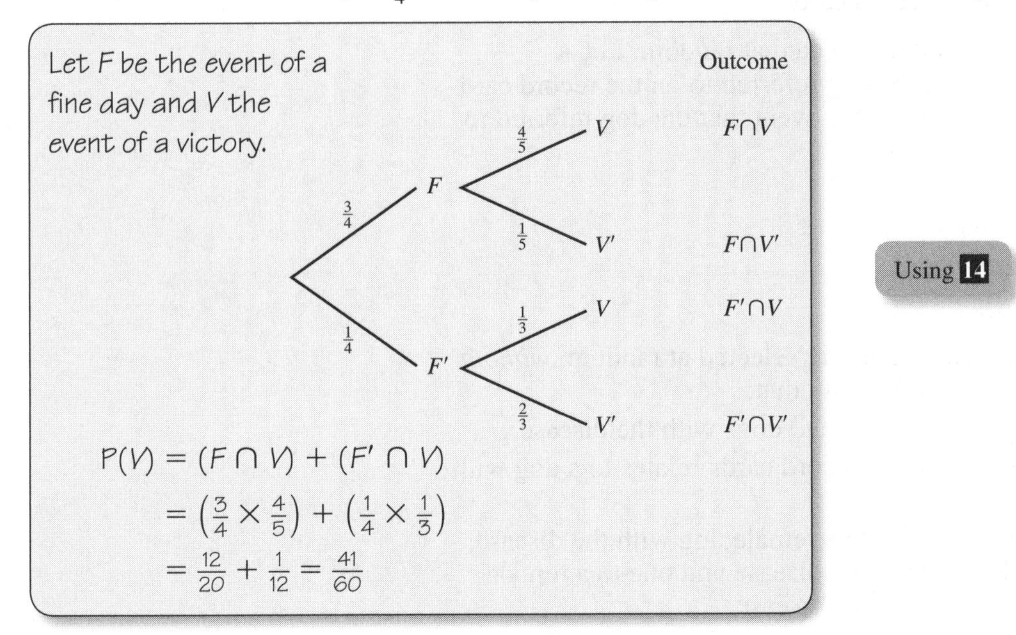

Using **14**

$$P(V) = (F \cap V) + (F' \cap V)$$
$$= \left(\tfrac{3}{4} \times \tfrac{4}{5}\right) + \left(\tfrac{1}{4} \times \tfrac{1}{3}\right)$$
$$= \tfrac{12}{20} + \tfrac{1}{12} = \tfrac{41}{60}$$

Worked examination question 1 [E]

The three events A, B and C are defined in the same sample space.
The events A and C are mutually exclusive. The events A and B are independent.
Given that $P(A) = \frac{3}{5}$, $P(C) = \frac{1}{3}$ and $P(A \cup B) = \frac{7}{8}$, find:

(a) $P(A \cup C)$ **(b)** $P(B)$.

(a) $P(A \cup C) = P(A) + P(C) - P(A \cap C)$

$$= \frac{3}{5} + \frac{1}{3} - 0$$
$$= \frac{9+5}{15} = \frac{14}{15}$$

Using **9**

Here we use **12**. As A and C are mutually exclusive $P(A \cap C) = 0$

(b) $P(A \cup B) = P(A) + P(B) - P(A \cap B)$

$$= P(A) + P(B) - P(A) \times P(B)$$
$$\frac{7}{8} = \frac{3}{5} + P(B) - \frac{3}{5} \times P(B)$$
$$\frac{7}{8} - \frac{3}{5} = \left(1 - \frac{3}{5}\right)P(B)$$
$$\frac{35-24}{40} = \frac{2}{5}P(B)$$
$$P(B) = \frac{11}{40} \times \frac{5}{2} = \frac{11}{16}$$

Using **9**

Here we use **11**. As A and B are independent. $P(A \cap B = P(A) \times P(B)$

Worked examination question 2 [E]

During 1996 a vet saw 125 dogs, each suspected of having a particular disease. Of the 125 dogs, 60 were female of whom 25 actually had the disease and 35 did not. Only 20 of the males had the disease, the rest did not. The case history of each dog was documented on a separate record card.

(a) A record card from 1996 is selected at random. Let A represent the event that the dog referred to on the record card was female and B represent the event that the dog referred to was suffering from the disease.

Find: **(i)** $P(A)$

(ii) $P(A \cup B)$

(iii) $P(A \cap B)$

(iv) $P(A | B)$.

(b) If three different record cards are selected at random, *without replacement*, find the probability that:

(i) all three record cards relate to dogs with the disease,

(ii) exactly one of the three record cards relates to a dog with the disease,

(iii) one record card relates to a female dog with the disease, one to a male dog with the disease and one to a female dog not suffering from the disease.

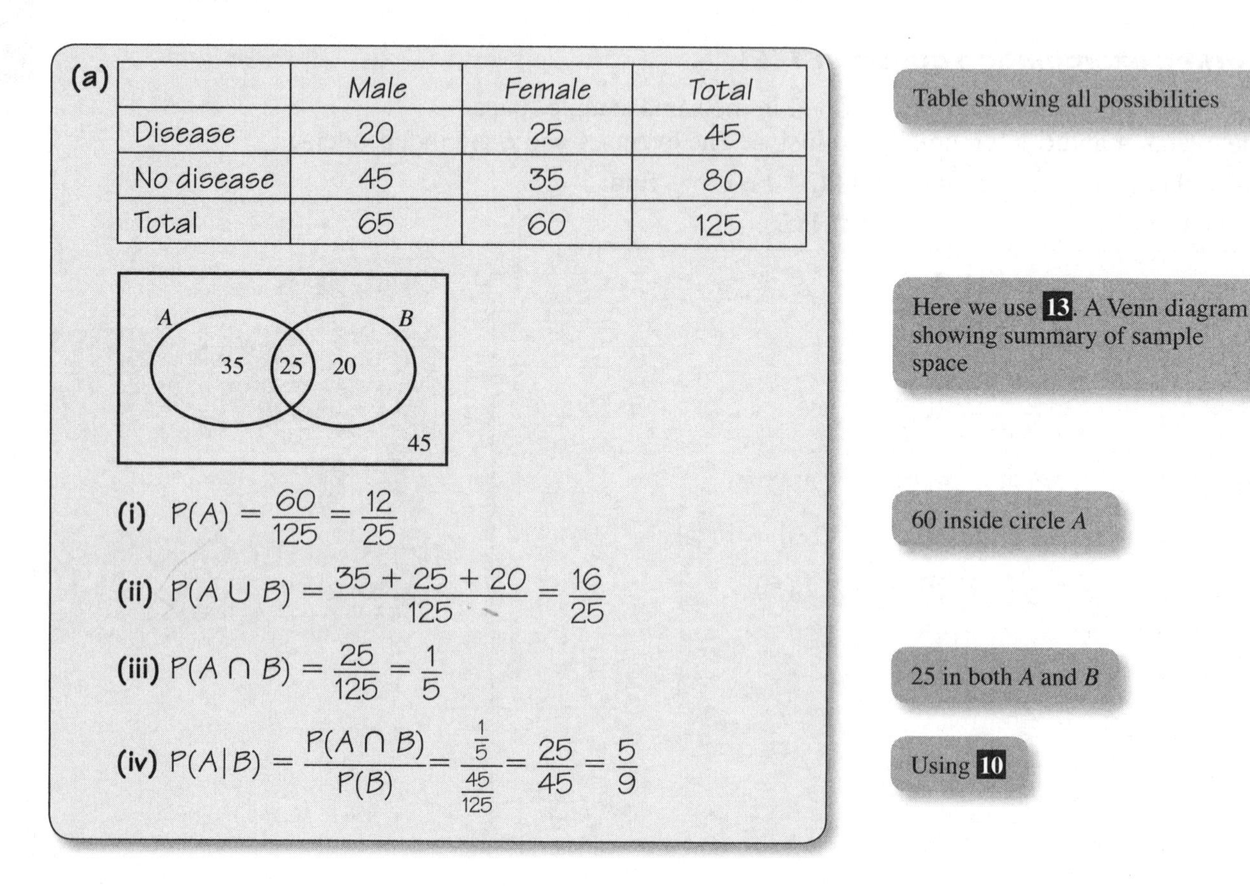

(a)

	Male	Female	Total
Disease	20	25	45
No disease	45	35	80
Total	65	60	125

Table showing all possibilities

Here we use **13**. A Venn diagram showing summary of sample space

(i) $P(A) = \frac{60}{125} = \frac{12}{25}$

60 inside circle A

(ii) $P(A \cup B) = \frac{35 + 25 + 20}{125} = \frac{16}{25}$

(iii) $P(A \cap B) = \frac{25}{125} = \frac{1}{5}$

25 in both A and B

(iv) $P(A|B) = \frac{P(A \cap B)}{P(B)} = \frac{\frac{1}{5}}{\frac{45}{125}} = \frac{25}{45} = \frac{5}{9}$

Using **10**

(b) Using **14**, Tree diagram

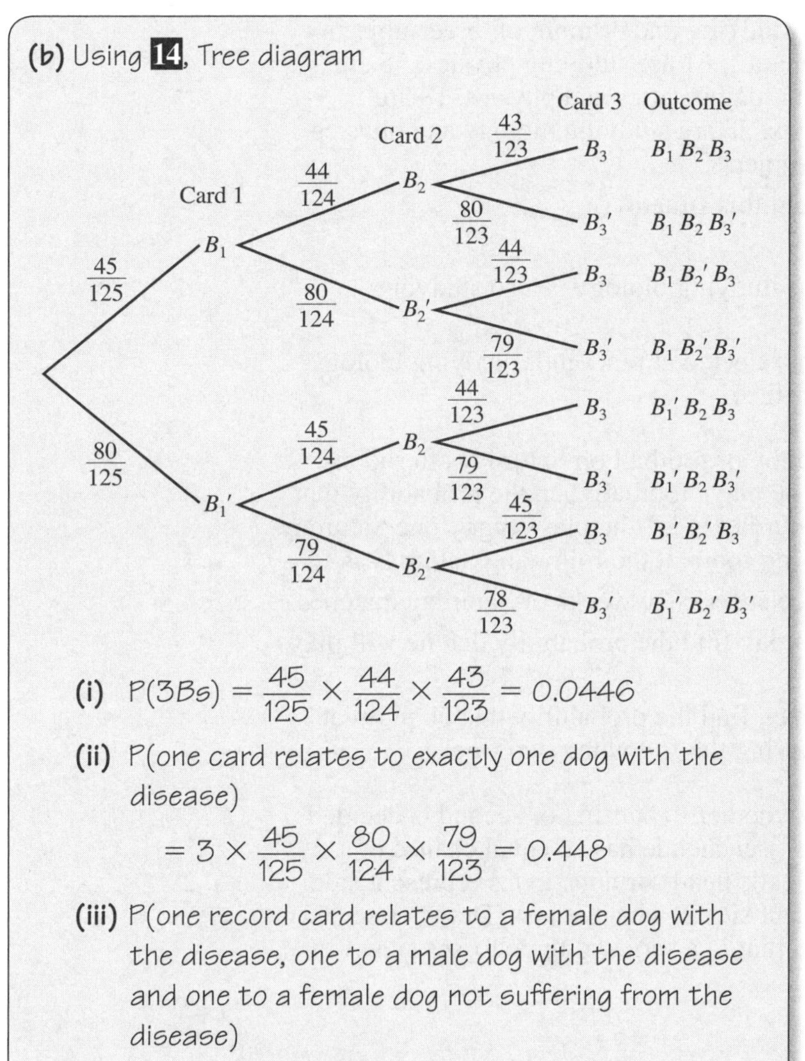

(i) $\text{P(3Bs)} = \frac{45}{125} \times \frac{44}{124} \times \frac{43}{123} = 0.0446$

(ii) P(one card relates to exactly one dog with the disease)

$$= 3 \times \frac{45}{125} \times \frac{80}{124} \times \frac{79}{123} = 0.448$$

(iii) P(one record card relates to a female dog with the disease, one to a male dog with the disease and one to a female dog not suffering from the disease)

$$= 6 \times \frac{25}{125} \times \frac{20}{124} \times \frac{35}{123} = 0.0551$$

Multiply along branches.

Possibilities are $B_1 B_2' B_3'$, $B_1' B_2 B_3'$ and $B_1' B_2' B_3$. Multiply along branches and add the results.

There are 6 ways in which three different cards can be drawn.

Revision exercise 5

1 State in words the relationship between two events A and B when:

(a) $P(A \cap B) = P(A) \times P(B)$

(b) $P(A \cap B) = 0$

2 The events A and B are independent and $P(A) = 0.4$ and $P(B) = 0.3$. Calculate:

(a) $P(A \cup B)$

(b) $P(A')$

3 In a sixth form 100 students are studying one or more of three subjects. 73 students are studying mathematics, 61 are studying physics, 35 are studying biology, 43 are studying mathematics and physics, 18 are studying mathematics and biology, 23 are studying biology and physics while 15 are studying all three subjects.

Draw a Venn diagram to represent this situation.

Use your diagram to find:

(a) the probability that a student studying biology is also studying mathematics but not physics

(b) the probability that a student selected at random is studying biology and physics but not mathematics.

4 A student always plays either rugby or football on Saturday afternoons, but never both. If one Saturday he plays football then the probability that he plays rugby the following Saturday is $\frac{3}{5}$. If he plays rugby one Saturday then the probability of him playing football the following Saturday is $\frac{2}{3}$.

(a) Draw a tree diagram for three successive weeks of Saturday matches.

(b) If he plays football one Saturday find the probability that he will play rugby in two weeks time.

(c) If he plays rugby one Saturday, find the probability that he plays at least one game of football during the following two weeks.

5 In cricket, the side that chooses whether to bat first or second is decided by the toss of a coin. Theoretically each side has an equal chance of winning the toss no matter who calls heads or tails. Let A represent side A wins the toss, and let B represent side B wins the toss. Draw a tree diagram and find the probability that in a series of three tosses side A wins the toss:

(a) three times

(b) exactly twice.

(c) Calculate the probability that side A wins the toss more times than it loses the toss.

(d) Comment on the fairness of this system.

6 Students in a class were given two statistics problems to solve. Within the class $\frac{5}{6}$ of the students solved the first one correctly and $\frac{7}{12}$ solved the second one correctly. Of those students who correctly solved the first problem, $\frac{3}{5}$ correctly solved the second one.

One student was chosen at random from the class.

Let A be the event that the student solved the first problem correctly and B the event that the student solved the second problem correctly.

(a) Find $\mathrm{P}(A \cap B)$ and $\mathrm{P}(A \cup B)$.

(b) Given that the student solved the second problem correctly find the probability that the first problem was solved correctly.

(c) Given that the student did not solve the second problem correctly find the probability that the first problem was correctly solved.

Correlation 6

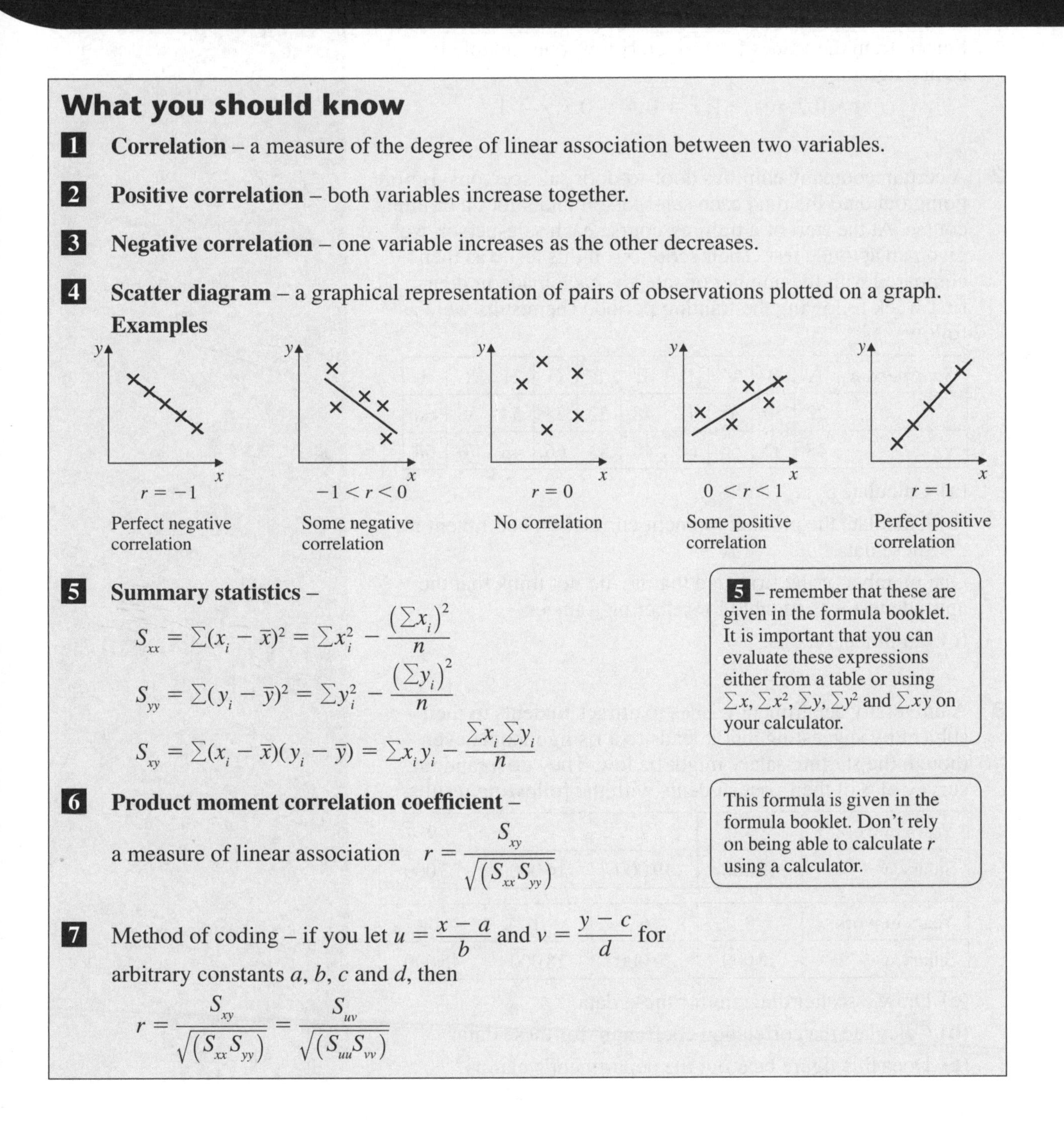

What you should know

1 **Correlation** – a measure of the degree of linear association between two variables.

2 **Positive correlation** – both variables increase together.

3 **Negative correlation** – one variable increases as the other decreases.

4 **Scatter diagram** – a graphical representation of pairs of observations plotted on a graph.

Examples

$r = -1$	$-1 < r < 0$	$r = 0$	$0 < r < 1$	$r = 1$
Perfect negative correlation	Some negative correlation	No correlation	Some positive correlation	Perfect positive correlation

5 **Summary statistics** –

$$S_{xx} = \sum(x_i - \bar{x})^2 = \sum x_i^2 - \frac{\left(\sum x_i\right)^2}{n}$$

$$S_{yy} = \sum(y_i - \bar{y})^2 = \sum y_i^2 - \frac{\left(\sum y_i\right)^2}{n}$$

$$S_{xy} = \sum(x_i - \bar{x})(y_i - \bar{y}) = \sum x_i y_i - \frac{\sum x_i \sum y_i}{n}$$

5 – remember that these are given in the formula booklet. It is important that you can evaluate these expressions either from a table or using $\sum x$, $\sum x^2$, $\sum y$, $\sum y^2$ and $\sum xy$ on your calculator.

6 **Product moment correlation coefficient** –

a measure of linear association $r = \dfrac{S_{xy}}{\sqrt{(S_{xx} S_{yy})}}$

This formula is given in the formula booklet. Don't rely on being able to calculate r using a calculator.

7 Method of coding – if you let $u = \dfrac{x - a}{b}$ and $v = \dfrac{y - c}{d}$ for arbitrary constants a, b, c and d, then

$$r = \frac{S_{xy}}{\sqrt{(S_{xx} S_{yy})}} = \frac{S_{uv}}{\sqrt{(S_{uu} S_{vv})}}$$

Test yourself

What to review

If your answer is incorrect:

1 Shown below are three different scatter diagrams:

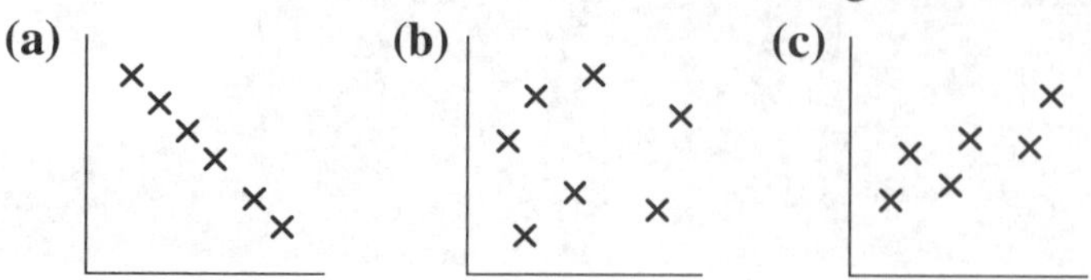

Select, from the values for r given below, one suitable for each diagram.

$$r = -0.7; r = -1; r = 0; r = 0.8; r = 1.$$

Review Edexcel Book S1 pages 115–118 and 124–125

2 A certain company employs door-to-door salespersons. Before going out onto the road each salesperson undergoes a training course. At the start of a training course each salesperson was given an aptitude test. Their score, x, on this test was then compared with the number of sales, y, each made in their first week following the training period. The results were as follows:

Salesperson	A	B	C	D	E	F	G	H	I	J
x	22	30	35	42	48	52	54	54	56	60
y	48	52	60	60	48	53	66	45	70	68

(a) Calculate S_{xx}, S_{yy} and S_{xy}.

(b) Calculate the product moment correlation coefficient for these data.

One member of the firm said that he did not think that the aptitude test was suitable for selecting trainees.

(c) Comment on this.

Review Edexcel Book S1 pages 119–125

3 A university department wishes to attract students to their subject by suggesting that it leads to a rising income even though the starting salary might be low. They do a random survey of 8 of their past students with the following results.

Years in work, x	4	7	5	9
Salary, y	12 000	19 000	16 000	23 000

Years in work, x	8	6	10	14
Salary, y	18 000	20 000	28 000	45 000

(a) Draw a scatter diagram for these data.

(b) Calculate the correlation coefficient for these data.

(c) Does this figure bear out the department's claims?

Review Edexcel Book S1 pages 115–125

Example 1

For ten cities the heights above sea level, x, in hundreds of metres, and their temperatures in degrees centigrade, y, on the same day in September were as follows:

Place	A	B	C	D	E	F	G	H	I	J
x	18	11	3	5	8	11	4	15	16	5
y	9	13	18	17	13	10	16	10	6	14

(a) Calculate S_{xx}, S_{yy}, and S_{xy} for these data.

(b) Calculate the product moment correlation coefficient for these data.

(c) Interpret your result.

(a) $\sum x_i = 96, \sum y_i = 126, \sum x_i^2 = 1186, \sum y_i^2 = 1720,$
$\sum x_i y_i = 1038.$

$$S_{xx} = \sum x_i^2 - \frac{(\sum x_i)^2}{n} = 1186 - \frac{96^2}{10}$$
$$= 1186 - 921.6 = 264.4$$

Using **5**

Note: If you have a suitable calculator you can arrive at the various totals and summary statistics without drawing up a table, so a table need not be given in answers to examination questions.

$$S_{yy} = \sum y_i^2 - \frac{(\sum y_i)^2}{n}$$
$$= 1720 - \frac{126^2}{10}$$
$$= 1720 - 1587.6 = 132.4$$

$$S_{xy} = \sum x_i y_i - \frac{\sum x_i \sum y_i}{n}$$
$$= 1038 - \frac{96 \times 126}{10}$$
$$= 1038 - 1209.6 = -171.6$$

(b) $$r = \frac{S_{xy}}{\sqrt{S_{xx} S_{yy}}} = \frac{-171.6}{\sqrt{(264.4 \times 132.4)}}$$
$$= -0.9172 = -0.917 \text{ (3 d.p.)}$$

Using **6**

(c) As height above sea level increases the temperature decreases.

Example 2

A group of sixth form students each had their height, h, in centimetres, and their weight, w, in kilograms measured. The resulting data are recorded below:

Student	A	B	C	D	E	F
h (cm)	176	170	180	175	189	178
w (kg)	79.1	76.3	85.5	79.2	77.5	90.2

Student	G	H	I	J	K	L
h (cm)	191	173	176	181	182	180
w (kg)	93.0	74.8	91.0	77.2	76.0	76.0

Use the coding $W = w - 80$ and $H = h - 175$.

(a) Draw a scatter diagram of W against H.

(b) Calculate the product moment correlation coefficient.

(c) Interpret your result.

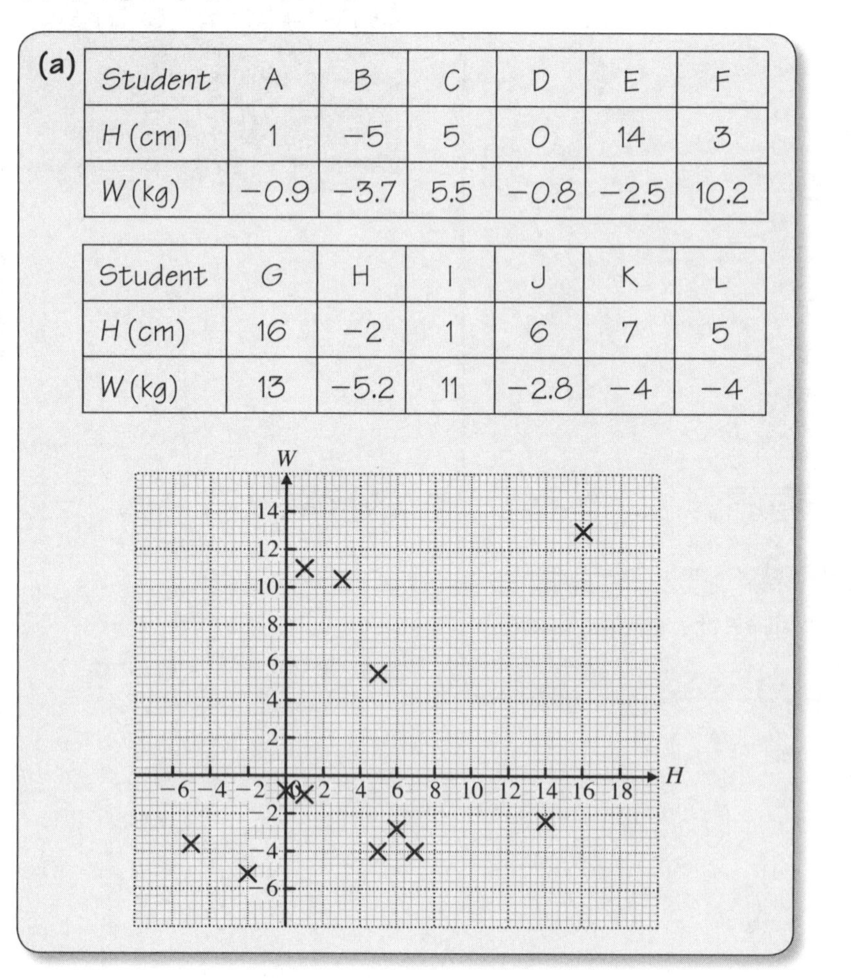
(a)

Student	A	B	C	D	E	F
H (cm)	1	−5	5	0	14	3
W (kg)	−0.9	−3.7	5.5	−0.8	−2.5	10.2

Student	G	H	I	J	K	L
H (cm)	16	−2	1	6	7	5
W (kg)	13	−5.2	11	−2.8	−4	−4

Here we use **7** and code the variables.

Using **4** and drawing a scatter diagram.

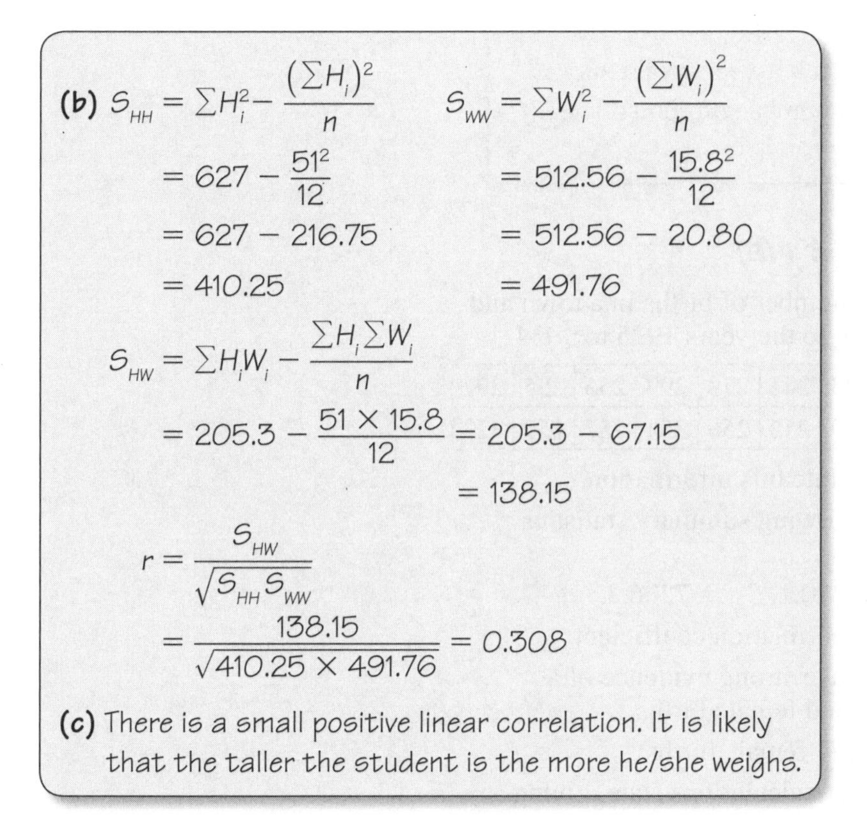

(b) $S_{HH} = \sum H_i^2 - \dfrac{(\sum H_i)^2}{n}$ $\qquad$ $S_{WW} = \sum W_i^2 - \dfrac{(\sum W_i)^2}{n}$

$= 627 - \dfrac{51^2}{12}$ $\qquad$ $= 512.56 - \dfrac{15.8^2}{12}$

$= 627 - 216.75$ $\qquad$ $= 512.56 - 20.80$

$= 410.25$ $\qquad$ $= 491.76$

$S_{HW} = \sum H_i W_i - \dfrac{\sum H_i \sum W_i}{n}$

$= 205.3 - \dfrac{51 \times 15.8}{12} = 205.3 - 67.15$

$= 138.15$

$r = \dfrac{S_{HW}}{\sqrt{S_{HH} S_{WW}}}$

$= \dfrac{138.15}{\sqrt{410.25 \times 491.76}} = 0.308$

(c) There is a small positive linear correlation. It is likely that the taller the student is the more he/she weighs.

Using **5**

Using **7**

Example 3

A keen football fan thought that the more goals his team scored during any given season, the higher the number of points they would score. Then he wondered what effect goals scored against his side would have. He obtained the results of all the matches for the particular league for a season and, using the number of points scored p and the goals scored against a, he calculated the following summary statistics: $n = 20$, $\sum a = 1073$, $\sum p = 801$, $\sum a^2 = 59\,923$, $\sum p^2 = 33\,619$, $\sum ap = 41\,433$.

Calculate the correlation between goals scored against and points scored and interpret the result.

$$S_{aa} = \sum a^2 - \frac{(\sum a)^2}{n} = 59\,923 - \frac{1073^2}{20} = 2356.55$$

$$S_{pp} = \sum p^2 - \frac{(\sum p)^2}{n} = 33\,619 - \frac{801^2}{20} = 1538.95$$

$$S_{ap} = \sum ap - \frac{\sum a \sum p}{n} = 41\,433 - \frac{1073 \times 801}{20}$$

$$= -1540.65$$

$$r = \frac{S_{ap}}{\sqrt{S_{aa} S_{pp}}} = \frac{-1540.65}{\sqrt{2356.55 \times 1538.95}} = -0.809$$

There is some negative correlation between goals scored against a football team and the number of points they get in a season.

Worked examination question 1 [E]

A local historian was studying the number of births in a town and found the following figures relating to the years 1925 to 1934.

Male births, x	223	218	223	223	242	278	299	256	255	292
Female births, y	219	205	209	239	252	256	254	257	259	323

(a) Draw a scatter diagram to illustrate this information.

The historian calculated the following summary statistics from the data:

$$S_{xx} = 8276.9, S_{yy} = 10\,230.1, S_{xy} = 7206.3.$$

(b) Calculate the product moment correlation coefficient,

The historian believed these data gave strong evidence of a positive correlation between male and female births.

In 1924 there were 249 male and 177 female births.

(c) Without carrying out any further calculations state, giving a reason, what effect the inclusion of these figures would have on the value of the product moment correlation coefficient.

(a) Using **4** a scatter diagram

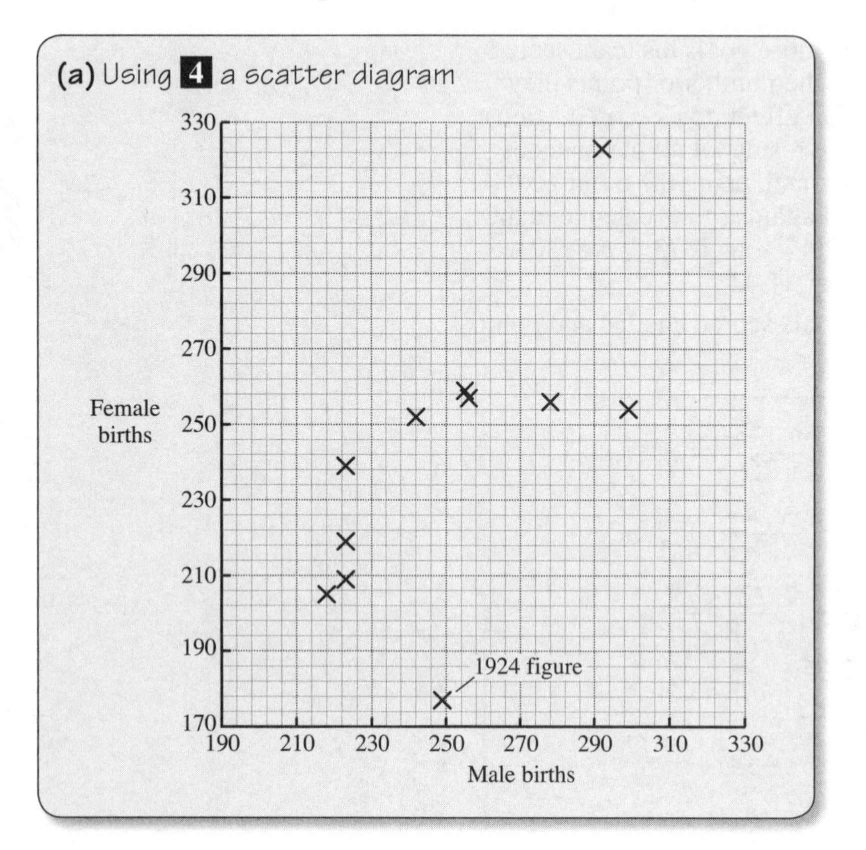

Using 6

(b) $r = \frac{S_{xy}}{\sqrt{S_{xx}S_{yy}}}$

$= \frac{7206.3}{\sqrt{8276.9 \times 10\,230.1}} = 0.783$

(c) Reduce r. From the scatter diagram the point would appear to be an outlier or away from the normal trend.

Revision exercise 6

1 'Tyger, tyger burning bright, in the forests of the night', is how William Blake begins his well-known poem. In the six verses of the poem the total number of words in each verse, y, and the number of words with three or fewer letters, x, are given in the table below.

Verse	1	2	3	4	5	6
x	7	7	9	7	11	7
y	20	25	27	23	28	20

(a) Calculate S_{xx}, S_{yy} and S_{xy} for these data.

(b) Calculate the value of the product moment correlation coefficient for this data.

(c) Comment on this value.

2 In certain commercial apple orchards apples are picked and graded. Then each apple has a sticky label attached showing the variety. To help with the ordering of labels it was thought that if there was a high positive correlation between weight of apples and number of labels, the labels could be ordered once an estimate of the weight was known. A trial was set up to look at the yields in kg, k, from orchards and the number of labels, l, used. A random sample from the last four years from the orchards owned by a grower gave the results shown in the table.

Orchard	A	B	C	D	E	F
Yield, k (kg)	8500	8850	8200	8750	8600	8580
Number of labels, l	68 000	75 000	67 000	73 500	70 500	71 000

You may assume that $\sum\left(\frac{k-8550}{10}\right) = 18$,

$\sum\left(\frac{k-8550}{10}\right)^2 = 2584$, $\sum\left(\frac{k-8550}{10}\right)\left(\frac{l-69\,000}{100}\right) = 3585$,

$\sum\left(\frac{l-69\,000}{100}\right) = 110$, $\sum\left(\frac{1-69\,000}{100}\right)^2 = 6750$

6

(a) Calculate the product moment correlation coefficient for the data.

(b) What conclusion would you draw from the answer to **(a)**?

3 Summary statistics for the number of actual submarines sunk, y, and the number reported sunk, x, each month for a 10-month period of the second world war were as given below.

$\sum x = 53$, $\sum y = 67$, $\sum x^2 = 381$, $\sum y^2 = 632$ and $\sum xy = 413$

(a) Calculate the correlation coefficient for these data.

(b) Does this mean that the reports were reasonably accurate?

Regression 7

What you should know

1 **Explanatory variable** – the variable (usually labelled X) whose values can be set accurately and which depend only on the experimenter's choice. Sometimes called the independent variable.

2 **Response variable** – the variable (usually labelled Y) whose values depend upon the values of the explanatory variable. Sometimes called the dependent variable.

3 **Scatter diagram** – the graph of the pairs of the two variables. The explanatory variable is plotted horizontally and the response variable is plotted vertically.

It is important that you do not mix up the explanatory variable with the response variable.

4 **Straight line law** – the law of a straight line is given as $y = a + bx$ where a and b are constants. The gradient of the line is b, and a is the intercept on the y-axis (value of y when $x = 0$).

5 **The equation of the least squares regression line of y on x** is given by

$$y = a + bx \text{ where } b = \frac{S_{xy}}{S_{xx}} \text{ and } a = \bar{y} - b\bar{x}.$$

The formulae for S_{xx} and S_{xy} are the same as those in the previous chapter and are given in the formulae booklet. The formula for the regression line is also given in the formula booklet.

6 **Interpolation** – using the least squares model to estimate the value of the response variable for a given value of the explanatory variable **within** the original range of values of the explanatory variable.

7 **Extrapolation** – using the least squares model to estimate the value of the response variable for a given value of the explanatory variable **outside** the original range of values of the explanatory variable. Extrapolation should be used with a great deal of caution and not at all if the chosen explanatory value is far outside the original range of values.

You could be asked to use interpolation and/or extrapolation to calculate other values of y for given values of x, or vice-versa. You could also be asked to justify or comment on the extrapolation process.

Test yourself

What to review

If your answer is incorrect:

1 A bar was simply supported at its ends in a horizontal position and various masses were hung from the mid-point of the bar. The deflection of the bar at the centre position was measured with the following results.

Mass, x kg	20	25	30	35	40	45	50
Deflection, y cm	0.20	0.32	0.34	0.40	0.49	0.59	0.65

Draw a scatter diagram for these data.

Review Edexcel Book S1 pages 115–117

2 In an experiment with simple pendulums a large weight was tied to a thin piece of wire. The length, l (in cm) of the wire was adjusted and the periodic time t (the time for one oscillation) was found. The results were as follows.

l cm	10	20	30	40	50
t seconds	0.6	0.9	1.1	1.3	1.4

(a) Calculate the equation of the linear regression line of t on l, and draw this line on a scatter diagram of the results.

(b) Calculate the periodic time of a pendulum of length 25 cm.

Review Edexcel Book S1 pages 136–139 and 142–144

3 The manager of a catering service found that the number, y, of loaves consumed at a party was related to the number, x, of people attending. The linear regression line between these two variables was of the form $y = a + bx$.

(a) What value would you expect a to have and why?

(b) What does the value of b represent?

Review Edexcel Book S1 pages 142–144

Example 1

In an experiment at a research institute the amount of water supplied in inches and the yield of a hay crop in tons per acre were recorded. The results are shown in the table below.

x (inches)	12	18	24	30	36	42	48
y (tons/acre)	5.3	5.7	6.3	7.2	8.0	8.7	8.4

(a) Assume that the relationship between the two variables is a linear one and find the least squares regression equation for these data.

(b) Use your data to calculate the likely yield for a year when the rainfall was: **(i)** 20 inches **(ii)** 6 inches.

(c) Give reasons why this second estimate should not be relied upon.

(a) The least squares regression equation is $y = a + bx$.

For the data given

$\sum x_i = 210, \sum y_i = 49.6, \sum x_i^2 = 7308, \sum x_i y_i = 1590.$

$$S_{xx} = \sum x_i^2 - \frac{(\sum x_i)^2}{n} = 7308 - \frac{210^2}{7} = 1008$$

$$S_{xy} = \sum x_i y_i - \frac{\sum x_i \sum y_i}{n} = 1590 - \frac{210 \times 49.6}{7} = 102$$

Using **5**,

$$b = \frac{S_{xy}}{S_{xx}} = \frac{102}{1008} = 0.1012 = 0.101 \text{ (3 d.p.)}$$

$$\text{and } a = \bar{y} - b\bar{x} = \frac{\sum y_i}{n} - 0.1012 \frac{\sum x_i}{n}$$

$$= \frac{49.6}{7} - 0.1012 \frac{210}{7}$$

$$= 7.086 - 3.036 = 4.05$$

The least squares equation is $y = 4.05 + 0.101x$

(b) When $x = 20$ inches

$$y = 4.05 + 0.1012 \times 20$$

$$= 6.074 \text{ tons/acre} = 6.07 \text{ tons/acre (3 d.p.)}$$

Using **6**

When $x = 6$ inches

$$y = 4.05 + 0.1012 \times 6$$

$$= 4.657 \text{ tons/acre} = 4.66 \text{ tons/acre (3 d.p.)}$$

Using **7**

(c) $x = 6$ inches is outside the range 12 to 48 inches and would not be reliable. For example at this level of rainfall the seeds might not even germinate.

7

Worked examination question 1 [E]

The government of a country considered making an investment to decrease the number of members of the population per doctor in order to try and reduce its infant mortality rate. (Infant mortality is measured as the number of infants per 1000 who die before reaching the age of 5.) A study was made of several other similar countries and the variables x, population per doctor, and y, infant mortality, were examined. The data are summarised by the following statistics:

$$\bar{x} = 440.57, \quad \bar{y} = 8.00, \quad S_{xy} = -1598.00, \quad S_{xx} = 174\,567.71$$

(a) Calculate the equation of the regression line of y on x.

(b) Given that the country at present has 380 people per doctor, estimate the infant mortality.

(c) Comment on the coefficient of x in the light of the government's plans.

(a) Using **5**, $b = \frac{S_{xy}}{S_{xx}} = \frac{-1598.00}{174\,567.71} = -0.009\,154$

$a = \bar{y} - b\bar{x}$

$= 8.00 + 0.009154 \times 440.57 = 12.03$

so $y = 12.0 - 0.009\,15x$

(b) If $x = 380$ then using **6**,

$y = 12.0 - 0.009\,15 \times 380 = 8.52$

(c) b is slightly negative $\Rightarrow$ Reducing x will increase infant mortality. The government's investment is not worthwhile.

Worked examination question 2 [E]

One measure of personal fitness is the time taken for an individual's pulse rate to return to normal after strenuous exercise; the greater the fitness, the shorter the time. Reg and Norman have the same normal pulse rates. Following a short programme of strenuous exercise they both recorded their pulse rates P at time t minutes after they had stopped exercising. Norman's results are given in the table below.

t	0.5	1.0	1.5	2.0	3.0	4.0	5.0
P	125	113	102	94	81	83	71

(a) Draw a scatter diagram to represent this information.

The equation of the regression line of P on t for Norman's data is $P = 122.3 - 11.0t$. Plot this on the scatter diagram.

(b) Use the above equation to estimate Norman's pulse rate 2.5 minutes after stopping the exercise programme.

Reg's pulse rate 2.5 minutes after stopping the exercise was 100.

The full data for Reg are summarised by the following statistics:

$$n = 8, \sum t = 19.5, \sum t^2 = 63.75, \sum P = 829, \sum Pt = 1867.$$

(c) Find the equation of the regression line of P on t for Reg's data.

(d) State, giving a reason, which of Reg or Norman you consider to be the fitter.

(a)

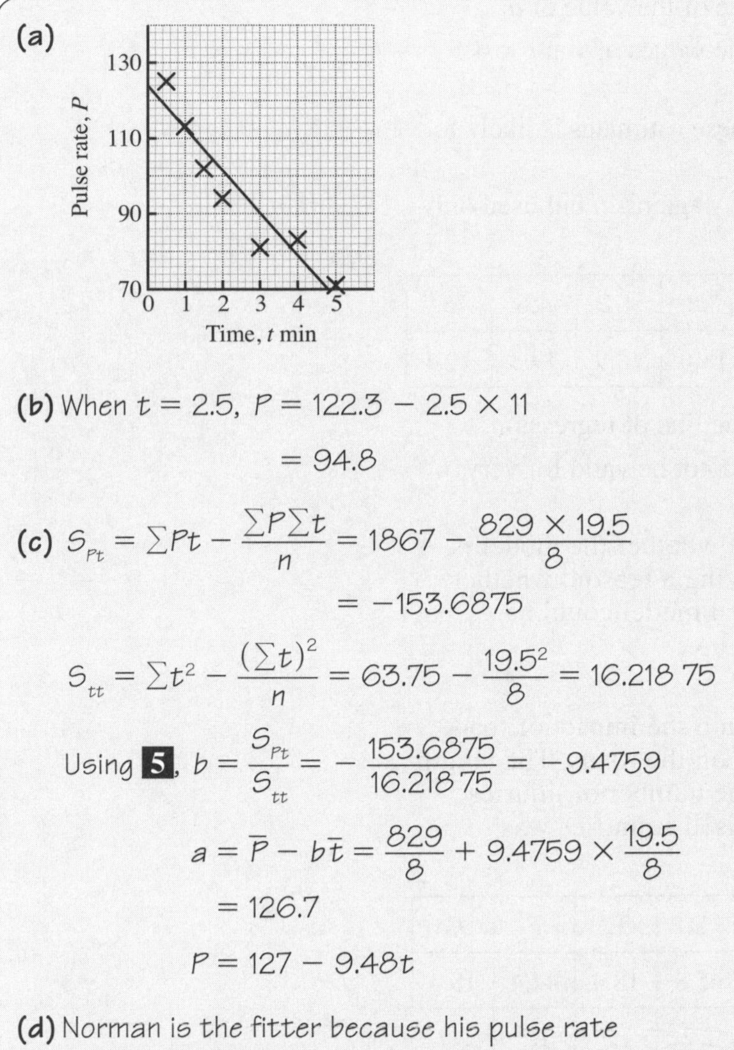

The regression line should always pass through the centre of the points on the diagram.

(b) When $t = 2.5$, $P = 122.3 - 2.5 \times 11$

$= 94.8$

(c) $S_{Pt} = \sum Pt - \dfrac{\sum P \sum t}{n} = 1867 - \dfrac{829 \times 19.5}{8}$

$= -153.6875$

$S_{tt} = \sum t^2 - \dfrac{(\sum t)^2}{n} = 63.75 - \dfrac{19.5^2}{8} = 16.218\,75$

Using **5**, $b = \dfrac{S_{Pt}}{S_{tt}} = -\dfrac{153.6875}{16.218\,75} = -9.4759$

Using **6**

$a = \bar{P} - b\bar{t} = \dfrac{829}{8} + 9.4759 \times \dfrac{19.5}{8}$

$= 126.7$

$P = 127 - 9.48t$

(d) Norman is the fitter because his pulse rate 2.5 minutes after stopping the exercise was lower than Reg's. This suggests a faster rate of return to normality, as does the steeper gradient of the regression line.

Revision exercise 7

1 In a physics experiment, a bottle of milk was brought from a cool room into a warm room. Its temperature, y °C, was recorded at t minutes after it was brought in, for 11 different values of t. The results are summarised as:

$$\sum t = 44,\ \sum t^2 = 180.4,\ \sum ty = 824.5,\ \sum y = 205.$$

(a) Calculate the equation of the line of regression of y on t in the form $y = a + bt$.

(b) Explain the practical significance of the value of a.

(c) Use your equation to estimate the values of y at $t = 4.5$ and $t = 20.0$.

(d) State, with a reason, which of these estimates is likely to be the more reliable.

The experimenter plotted a graph of y against t, but used only the data in the table below.

Time (minutes), t	3	3.4	3.8	4.2	4.6	5
Temperature (°C), y	17	18.3	18.6	18.9	19.3	19.4

(e) Plot this graph, and on it draw the line of regression.

(f) State why the linear model could not be valid for very large values of time.

(g) Using your graph, comment on whether the model is a reasonable one, and state, giving a reason, whether you consider that a more refined model could be found. [E]

2 An investigation was undertaken into the impact of road building on the number of injuries on the roads. The length of main road u (in 1000 km) and the number of injuries y (in 10 000) for several industrialised countries was recorded.

Country	A	B	C	D	E	F	G
Main roads, u	12.4	28.5	31.2	45.8	18.4	44.4	18.4
Injuries, y	3.1	2.4	4.5	2.2	1.7	1.1	2.7

(a) Plot a scatter diagram to illustrate this information.

The investigators suggested that a model of the form $y = p + qu$, where p and q are constants, might be suitable to describe the relationship between y and u.

(b) Comment briefly on this suggestion. (No calculations are required.)

A further investigation was undertaken into the impact of motorways on the number of injuries. The information showing the percentage, x, of main roads that were of motorway standard is given below.

Country	A	B	C	D	E	F	G
x	24.2	23.2	27.6	13.1	11.4	4.7	8.2

(c) Plot a scatter diagram of y against x.

(d) Obtain the equation of the regression line of y on x in the form $y = a + bx$.

[You may use $\sum x^2 = 2276.54$ and $\sum xy = 330.41$.]

(e) Give a practical interpretation of the constant a.

(f) Draw your regression line on the scatter diagram drawn in part **(c)**.

(g) Discuss briefly whether a straight line appears to be a suitable model for the relationship between y and x.

Country E is planning to upgrade a large number of its main roads to motorways.

(h) Explain briefly what the implications of the above investigations are on their plans. [E]

3 The height of a seedling, y millimetres, x weeks after it was planted is given in the following table.

x	5	6	7	8	9	10
y	102	111	123	135	148	153

(a) Plot a scatter diagram of y against x.

(b) Given that $\sum x = 45$, $\sum y = 772$, $\sum x^2 = 355$ and $\sum xy = 5979$, calculate the equation of the regression line for y on x in the form $y = a + bx$.

(c) What does the constant a represent?

(d) Calculate the expected height of the seedling after 12 weeks.

(e) How reliable is this estimate?

7

Discrete random variables

8

What you should know

1 **Random variable X** – a variable that represents the values obtained when we take a measurement from an experiment in the real world.

> **1** A random variable is often defined as x = number of … . You do not have to use X. As with events, it is worth trying to link the letter with the meaning e.g. G = the number of girls in a family of 4.

2 **Discrete random variable** – a variable that changes by steps, and takes only specified values in any given interval.

3 **Probability function** – a function that describes how the probabilities are assigned.

4 **Probability distribution** – the set of all values of a random variable together with their associated probabilities. For example:

m:	1	2	3	4	5	6
$P(M = m)$:	$\frac{1}{6}$	$\frac{1}{6}$	$\frac{1}{6}$	$\frac{1}{6}$	$\frac{1}{6}$	$\frac{1}{6}$

is the probability distribution which describes the values, m, taken when a die is thrown and the associated probabilities. Note $P(M = m)$ is sometimes written as $p(m)$.

5 **Sum of all probabilities** $\sum_{\forall x} p(x) = 1$

> **5** This is an important property. If you have to calculate probabilities always check that your probabilities add up to 1. See example 3.

6 **Cumulative distribution function (c.d.f.)**

$$F(x_0) = P(X \leqslant x_0) = \sum_{x \leqslant x_0} p(x)$$

7 **Expected value of X** $\quad E(X) = \mu = \sum_{\forall x} x P(X = x)$.

8 **Variance of X**

$$\text{Var}(X) = \sigma^2 = \sum_{\forall x} (x - \mu)^2 \, P(X = x).$$

$$= \sum_{\forall x} x^2 P(X = x) - \mu^2 = E(X^2) - \mu^2.$$

> **8** $\sigma^2 = \sum x^2 P(X = x) - \mu^2$ is usually the best formula to use, but don't forget the $-\ \mu^2$!

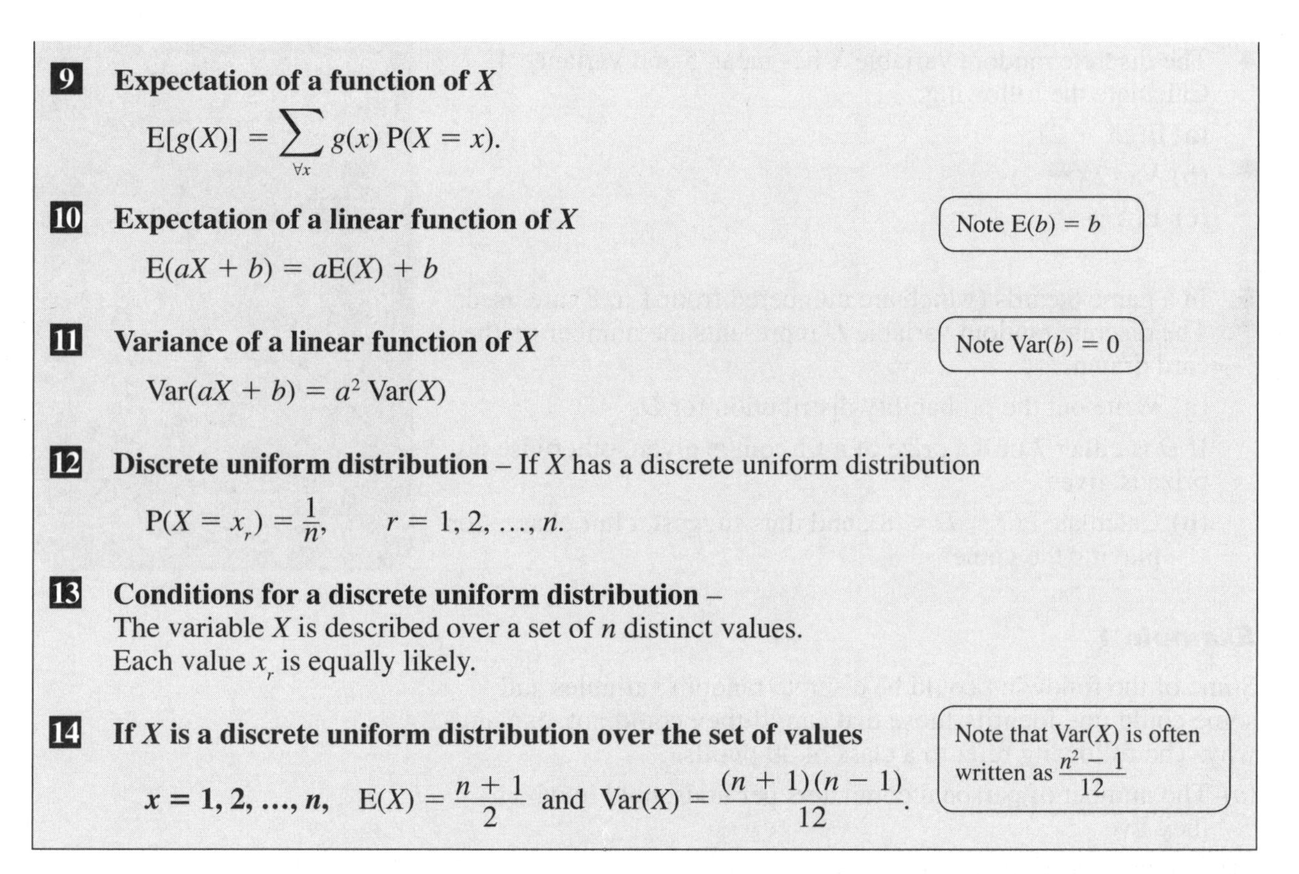

9 **Expectation of a function of X**

$$E[g(X)] = \sum_{\forall x} g(x)\, P(X = x).$$

10 **Expectation of a linear function of X**

$$E(aX + b) = aE(X) + b$$

Note $E(b) = b$

11 **Variance of a linear function of X**

$$Var(aX + b) = a^2\, Var(X)$$

Note $Var(b) = 0$

12 **Discrete uniform distribution** – If X has a discrete uniform distribution

$$P(X = x_r) = \frac{1}{n}, \qquad r = 1, 2, \ldots, n.$$

13 **Conditions for a discrete uniform distribution** –
The variable X is described over a set of n distinct values.
Each value x_r is equally likely.

14 **If X is a discrete uniform distribution over the set of values**

$$x = 1, 2, \ldots, n, \quad E(X) = \frac{n+1}{2} \text{ and } Var(X) = \frac{(n+1)(n-1)}{12}.$$

Note that $Var(X)$ is often written as $\frac{n^2 - 1}{12}$

Test yourself

What to review

If your answer is incorrect:

1 Find the mean and variance of the following distribution:

x	-3	-2	-1	0	1
$P(X = x)$	$\frac{1}{10}$	$\frac{1}{5}$	$\frac{2}{5}$	$\frac{1}{5}$	$\frac{1}{10}$

Review Edexcel Book S1 pages 159–164

2 The variable X has a c.d.f. $F(x) = \frac{3x - 1}{8}$; $x = 1, 2, 3$.

(a) Find $F(3)$.

(b) Show that $P(X = 2) = \frac{3}{8}$.

(c) Write down the probability distribution of X.

Review Edexcel Book S1 pages 156–158

3 There are two bags. Bag one contains 5 red counters numbered from 1 to 5, and bag two contains 4 blue counters numbered from 1 to 4. A counter is drawn from each bag and S represents the sum of their scores.

(a) Find the distribution of S.

(b) Calculate $E(S)$.

(c) Calculate $Var(S)$.

Review Edexcel Book S1 pages 152–154 and 159–164

4 The discrete random variable X has mean 5 and variance 4. Calculate the following:

(a) $E(2X + 2)$

(b) $Var(2X + 2)$

(c) $E(X^2)$

Review Edexcel Book S1 pages 165–168

5 In a game 8 cards (which are numbered from 1 to 8) are used. The discrete random variable D represents the number on the card drawn.

(a) Write out the probability distribution for D.

If D is either 7 or 8 a prize of a £1 coin is given, otherwise no prize is given.

(b) Calculate $P(7 \leqslant D \leqslant 8)$, and thus suggest a fair charge for playing the game.

Review Edexcel Book S1 pages 170–171

Example 1

Some of the following could be discrete random variables and some could not. Identify those that can. If they could not, explain why. The following refer to a class of 30 pupils.

(a) The number of personal computers per household in which they live.

(b) The heights of the pupils.

(c) The number of left-handed pupils.

(d) The number of cups of tea or coffee that they drink per day.

(e) The colour of the jumpers they are wearing.

Using **2**,

(a), **(c)** and **(d)** are discrete random variables.

(b) Cannot be a discrete random variable because height is a continuous variable.

(e) Cannot be a discrete random variable because colour is a qualitative variable and not a number.

Example 2

A discrete random variable x has the probability distribution given by:

x:	1	2	3	4
$P(X = x)$:	$\frac{1}{10}$	$\frac{1}{5}$	$\frac{k}{10}$	$\frac{3}{10}$

where k is a constant.

(a) Find the value of k.

(b) Find $P(X > 2)$.

(a) Using **5**, $\sum_{\forall x} p(x) = 1 \Rightarrow \frac{1}{10} + \frac{1}{5} + \frac{k}{10} + \frac{3}{10} = 1$

$$\frac{1 + 2 + k + 3}{10} = \frac{10}{10}$$

$$1 + 2 + k + 3 = 10$$

$$k = 10 - 6 = 4$$

(b) Using **4**, If $X > 2$ then $x = 3$ or 4.

$$P(X > 2) = P(X = 3) + P(X = 4) = \frac{4}{10} + \frac{3}{10} = \frac{7}{10}$$

Example 3

A discrete random variable X has cumulative distribution function $F(x)$ defined by:

$$F(x) = \frac{x + 4}{8}; x = 0, 1, 2, 3, 4.$$

Calculate:

(a) F(3) **(b)** $P(X \leqslant 2)$.

(c) Find the probability distribution of X.

(a) Using **6**, $F(3) = \frac{3 + 4}{8} = \frac{7}{8}$

(b) $P(X \leqslant 2) = F(2) = \frac{2 + 4}{8} = \frac{6}{8} = \frac{3}{4}$

(c) You can calculate $F(x)$ for $x = 0, 1$ and 4 in the same way to get $\frac{1}{2}, \frac{5}{8}$ and 1 respectively.

$$P(X = 1) = P(X \leqslant 1) - P(X \leqslant 0) = \frac{5}{8} - \frac{1}{2} = \frac{1}{8}$$

$$P(X = 2) = P(X \leqslant 2) - P(X \leqslant 1) = \frac{3}{4} - \frac{5}{8} = \frac{1}{8}$$

$$P(X = 3) = P(X \leqslant 3) - P(X \leqslant 2) = \frac{7}{8} - \frac{3}{4} = \frac{1}{8}$$

$$P(X = 4) = P(X \leqslant 4) - P(X \leqslant 3) = 1 - \frac{7}{8} = \frac{1}{8}$$

Using definition **4**,

the p.d.f. of x is:

x:	0	1	2	3	4
$P(X = x)$:	$\frac{1}{2}$	$\frac{1}{8}$	$\frac{1}{8}$	$\frac{1}{8}$	$\frac{1}{8}$

Check that $\sum_{\forall x} p(x) = 1$

$\frac{1}{2} + \frac{1}{8} + \frac{1}{8} + \frac{1}{8} + \frac{1}{8} = 1$ so the answer is correct.

8

Worked examination question 1 [E]

The discrete random variable X has the following probability distribution.

x	0	1	2	3	4
$P(X = x)$	0.2	0.2	0.2	0.2	0.2

(a) Write down the name of the distribution of X.

(b) Find $P(0 \leqslant X \leqslant 2)$.

(c) Find $E(X)$.

(d) Find $\text{Var}(X)$.

(e) Find $E(X^2 + 3)$

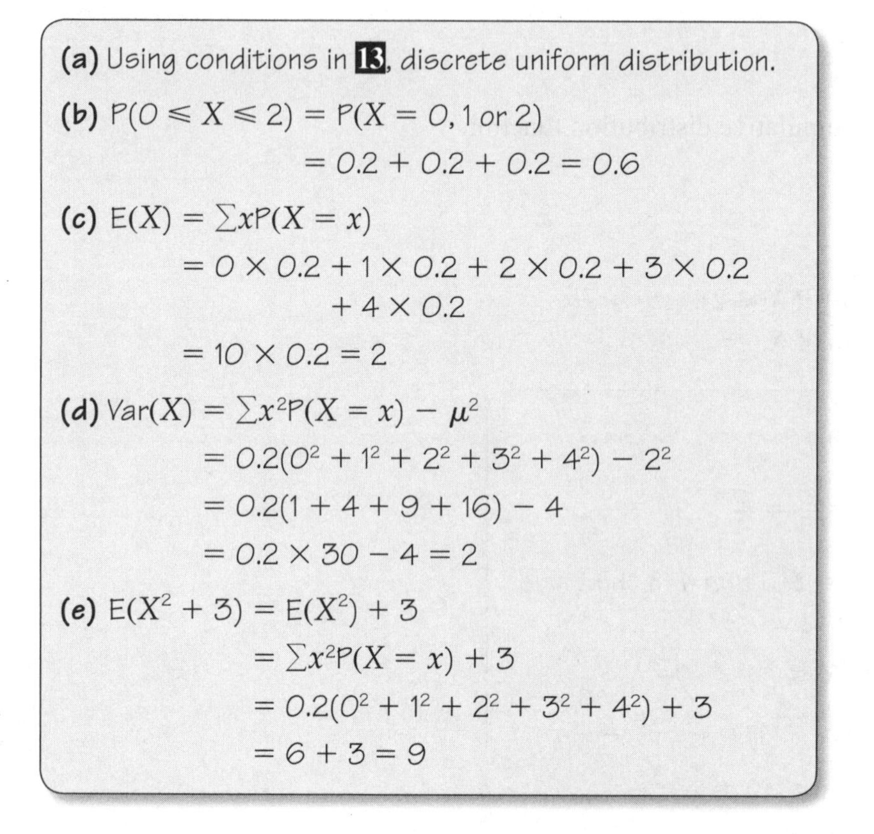

(a) Using conditions in **13**, discrete uniform distribution.

(b) $P(0 \leqslant X \leqslant 2) = P(X = 0, 1 \text{ or } 2)$

$= 0.2 + 0.2 + 0.2 = 0.6$

(c) $E(X) = \sum xP(X = x)$

$= 0 \times 0.2 + 1 \times 0.2 + 2 \times 0.2 + 3 \times 0.2 + 4 \times 0.2$

$= 10 \times 0.2 = 2$

(d) $\text{Var}(X) = \sum x^2 P(X = x) - \mu^2$

$= 0.2(0^2 + 1^2 + 2^2 + 3^2 + 4^2) - 2^2$

$= 0.2(1 + 4 + 9 + 16) - 4$

$= 0.2 \times 30 - 4 = 2$

(e) $E(X^2 + 3) = E(X^2) + 3$

$= \sum x^2 P(X = x) + 3$

$= 0.2(0^2 + 1^2 + 2^2 + 3^2 + 4^2) + 3$

$= 6 + 3 = 9$

Using **4**

Using **7**

Using **8**

Using **10**

Using **9**

Alternatively $E(X^2) = \text{Var}(X) + \mu^2$ could be used to find the $E(X^2)$ part.

Worked examination question 2 [E]

A fair cubical die has two yellow faces and four blue faces. The die is rolled repeatedly until a yellow face appears uppermost or the die has been rolled four times. The random variable B represents the number of times a blue face appears uppermost and the random variable R represents the number of times the die is rolled.

(a) Show that $P(B = 3) = \frac{8}{81}$.

(b) Find the probability distribution of B.

(c) Find $E(B)$.

(d) Show that $P(R = 4) = \frac{8}{27}$.

(e) Find $P(R = B)$.

(a) P(Yellow) $= \frac{1}{3}$, P(Blue) $= \frac{2}{3}$.

$P(B = 3) \Rightarrow BBBY$

$\therefore \quad P(B = 3) = \left(\frac{2}{3}\right)^3 \times \frac{1}{3} = \frac{8}{81}$

(b)

Outcome:	Y	BY	BBY	BBBY	BBBB
b:	0	1	2	3	4
$P(B = b)$:	$\frac{27}{81}$	$\frac{18}{81}$	$\frac{12}{81}$	$\frac{8}{81}$	$\frac{16}{81}$

(c) $E(B) = \frac{1}{81}\{0 + 18 + 24 + 24 + 64\} = \frac{130}{81} = \frac{149}{81}$

(d) $R = 4 \Rightarrow$ BBBY and BBBB

$\therefore \quad P(R = 4) = \frac{8}{81} + \frac{16}{81} = \frac{24}{81} = \frac{8}{27}$

(e) $R = B \Rightarrow BBBB$

$\therefore \quad P(R = B) = \frac{16}{81}$

The events are independent so $P(BBBY) = P(B)^3P(Y)$

Here we use **4**. Each $P(B = b)$ is calculated as in part **(a)**.

By **7**, $E(B) = \sum bP(B = b)$

$R = B$ only if no yellow is thrown.

Revision exercise 8

1 The discrete random variable x has the probability function shown in the table below.

x	1	2	3	4	5
$P(X = x)$	0.1	0.2	0.4	0.2	0.1

Calculate: **(a)** $P(2 < X \leqslant 4)$ **(b)** $F(3.7)$ **(c)** $E(X)$ **(d)** $Var(X)$ **(e)** $E(2X^2 - 3)$.

2 Given that the random variable X is the number showing when a die is thrown, and that the die is unbiased:

(a) write down the name of the probability distribution of X

(b) write down the cumulative distribution function of X.

(c) Using your answer to part **(b)** calculate:
(i) $P(X < 4)$ **(ii)** $P(2 < X \leqslant 5)$.

(d) Calculate:
(i) $E(X)$ **(ii)** $Var(X)$.

3 The discrete random variable X has the probability function shown in the table:

x	-1	0	1	2	3
$P(X = x)$	$\frac{1}{10}$	$\frac{1}{5}$	$\frac{1}{5}$	$\frac{3}{10}$	$\frac{1}{5}$

Calculate: **(a)** $P(-1 \leqslant X \leqslant 1)$ **(b)** $E(X)$ **(c)** $E(X^2)$ **(d)** $Var(X)$.

You can use **14** here.

8

4 The discrete random variable X has a c.d.f. given by

$$F(x) = \frac{x^2 + 1}{10}; x = 0, 1, 2, 3.$$

(a) Calculate F(0), F(1), F(2) and F(3).

(b) Write down the p.d.f. of X.

5 When a certain type of cell is subjected to radiation, the cell may die, survive as a single cell, or divide into two cells with probabilities $\frac{1}{2}, \frac{1}{3}, \frac{1}{6}$ respectively.

Two cells are independently subjected to radiation. The random variable X represents the total number of cells in existence after this experiment.

(a) Show that $P(X = 2) = \frac{5}{18}$.

(b) Find the probability distribution for X.

(c) Evaluate E(X).

(d) Show that $\text{Var}(X) = \frac{10}{9}$. [E]

6 An economics student is trying to model the daily movement, X points, in a stock market indicator. The student assumes that the value of X on one day is independent of the value on the next day. A fair die is rolled and if an odd number is uppermost then the indicator is moved down that number of points. If an even number is uppermost then the indicator is moved up that number of points.

(a) Write down the distribution of X as specified by the student's model.

(b) Find the value of E(X).

(c) Show that $\text{Var}(X) = \frac{179}{12}$.

If the indicator moves upwards over a period of time then this is taken as a sign of growth in the economy, if it falls then this is a sign that the economy is in decline.

(d) Comment on the state of the economy as suggested by this model.

Before the stock market opened one Monday morning the economic indicator was 3373.

(e) Use the student's model to find the probability that the indicator is at least 3400 when the stock market closes on the Friday afternoon of the same week. [E]

The normal distribution 9

What you should know

1 **The normal distribution** – is symmetrical about its mean μ
- has mode, median and mean all equal
- ranges from $-\infty$ to $+\infty$
- is asymptotic to the horizontal axis as $x \to -\infty$ and $x \to +\infty$
- has a total area under the curve of unity
- has notation $N(\mu, \sigma^2)$ where μ is the mean and σ^2 the variance of the distribution

2 **The standard normal distribution** – the standard normal variable is denoted by Z where $Z \sim N(0, 1^2)$
- any normal variable X, having mean μ and variance σ^2, can be transformed into a standard normal variable using

$\frac{X - \mu}{\sigma}$. Always remember that the denominator is σ.

3 $\Phi(z)$ – represents the area to the left of any given value z.

$$\Phi(z) = P(Z \leqslant z)$$

When z is negative:

$$\Phi(z) = 1 - \Phi(-z)$$

4

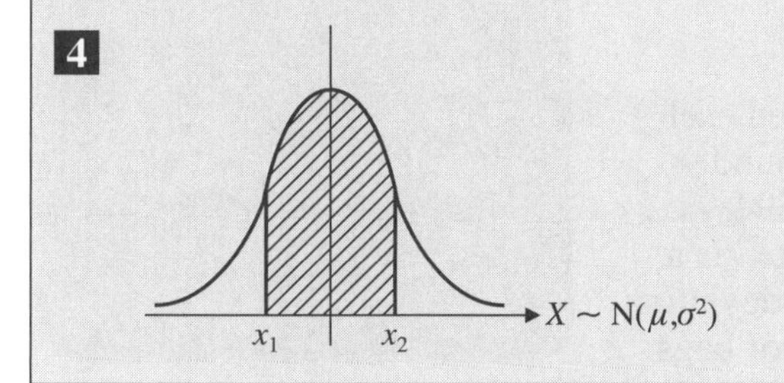

A good clear diagram illustrating the area corresponding to the required probability should always be drawn. In the example the shaded area shows $P(x_1 < X < x_2)$.

All the examples and exercises in this chapter using the normal distribution function will be based on z-values which have been evaluated to 2 decimal places. In the S1 examination, interpolation will be accepted although it is clearly stated in the specification for S1 that it is not necessary. The use of a calculator that can evaluate areas under the normal curve will also be acceptable.

Test yourself

What to review

If your answer is incorrect:

1 The standard normal variable $Z \sim N(0, 1^2)$.

Find:

(a) $P(Z < 1.43)$

(b) $P(Z < -1.77)$

(c) $P(-1.51 < Z < 1.33)$.

Review Edexcel Book S1 pages 177–178

2 The random variable X is distributed normally with mean 76 and variance 14.

Find:

(a) $P(X < 74.2)$

(b) $P(X > 81.2)$

(c) $P(73.8 < X < 82.1)$.

Review Edexcel Book S1 pages 182–183

3 The distance a particular car is able to travel on a full tank of fuel can be modelled by a normal random variable D with mean 640 km and standard deviation 13.5 km. A driver sets off in the car with a full tank of fuel on a journey of d km. Find, to the nearest km, the value of d such that there is no more than a 5% chance that the car runs out of fuel.

Review Edexcel Book S1 pages 179–184

4 On each occasion, over a period of months, that a particular repair has been carried out by an engineer the time taken, T minutes, has been recorded. It is assumed that T can be modelled as a normal random variable with mean 155 minutes.

(a) Given that $P(T < 161) = 0.8849$, calculate the standard deviation of T.

(b) Find $P(146 < T < 166)$.

Review Edexcel Book S1 pages 185–188

5 A builder is reviewing the amount of cement C kg used each month by his employees. From past records he has found that $P(C \geqslant 3390) = 0.7123$ and $P(C \geqslant 3675) = 0.0694$. The cement is delivered in 25 kg bags. Assuming that C is a normal random variable, find the mean and standard deviation of C. Give your answer to the nearest whole number of bags.

Review Edexcel Book S1 pages 185–188

Example 1

The standard normal variable $Z \sim N(0, 1^2)$.
Find:

(a) $P(Z < 1.69)$ **(b)** $P(Z < -2.18)$ **(c)** $P(-1.31 < Z < 1.06)$

Note that even in the simplest cases such as this it is always advisable to draw a diagram.

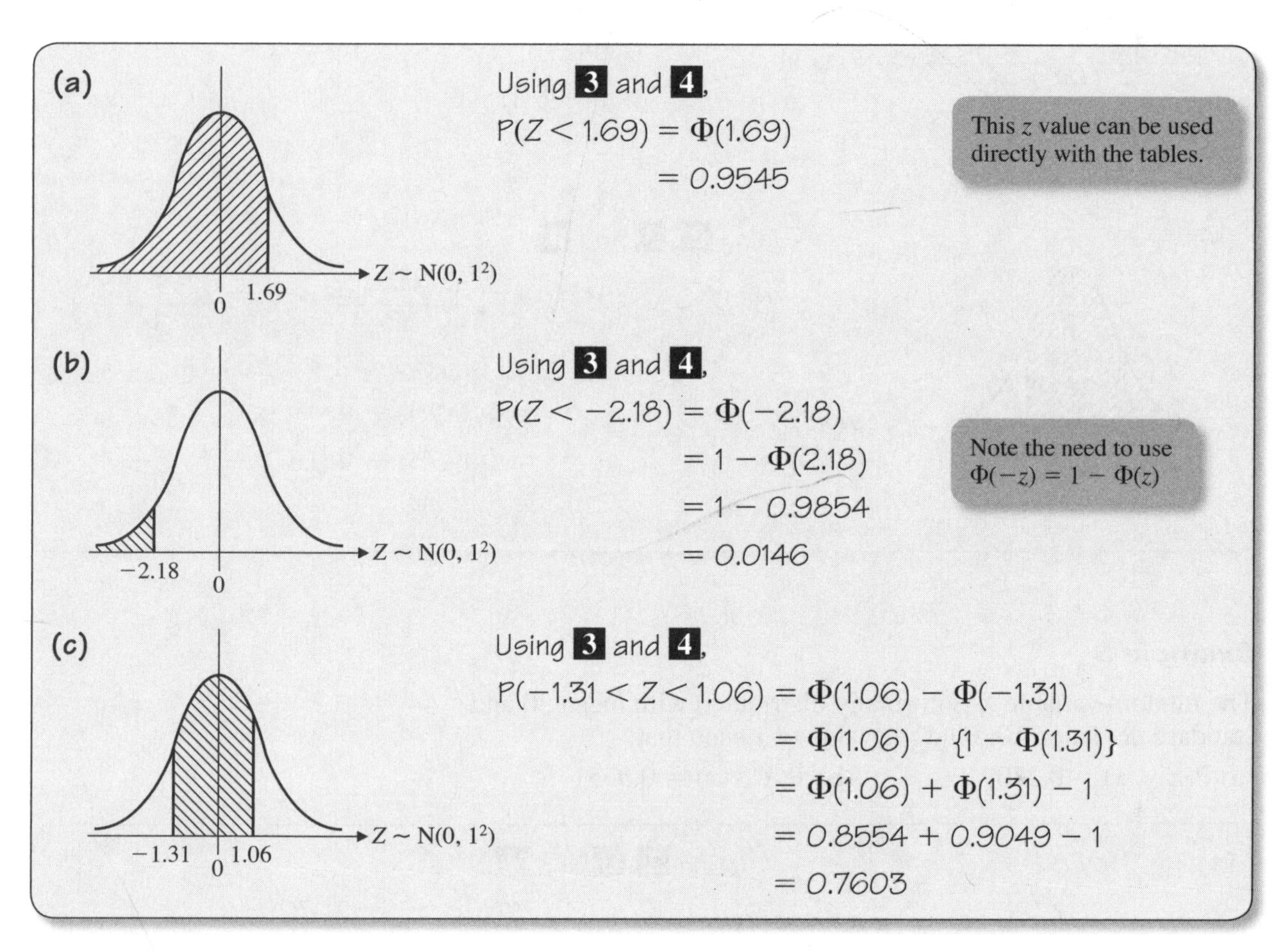

(a) Using **3** and **4**,

$P(Z < 1.69) = \Phi(1.69)$
$= 0.9545$

This z value can be used directly with the tables.

(b) Using **3** and **4**,

$P(Z < -2.18) = \Phi(-2.18)$
$= 1 - \Phi(2.18)$
$= 1 - 0.9854$
$= 0.0146$

Note the need to use $\Phi(-z) = 1 - \Phi(z)$

(c) Using **3** and **4**,

$P(-1.31 < Z < 1.06) = \Phi(1.06) - \Phi(-1.31)$
$= \Phi(1.06) - \{1 - \Phi(1.31)\}$
$= \Phi(1.06) + \Phi(1.31) - 1$
$= 0.8554 + 0.9049 - 1$
$= 0.7603$

Example 2

The random variable x is distributed normally with mean 120 and variance 64.
Find:

(a) $P(X > 126)$ **(b)** $P(X < 112)$ **(c)** $P(108 < X < 126)$

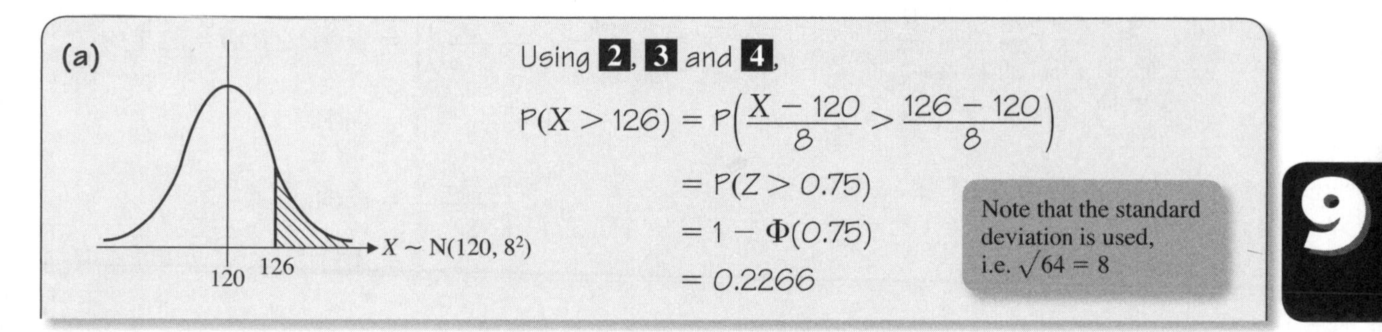

(a) Using **2**, **3** and **4**,

$$P(X > 126) = P\left(\frac{X - 120}{8} > \frac{126 - 120}{8}\right)$$
$= P(Z > 0.75)$
$= 1 - \Phi(0.75)$
$= 0.2266$

Note that the standard deviation is used, i.e. $\sqrt{64} = 8$

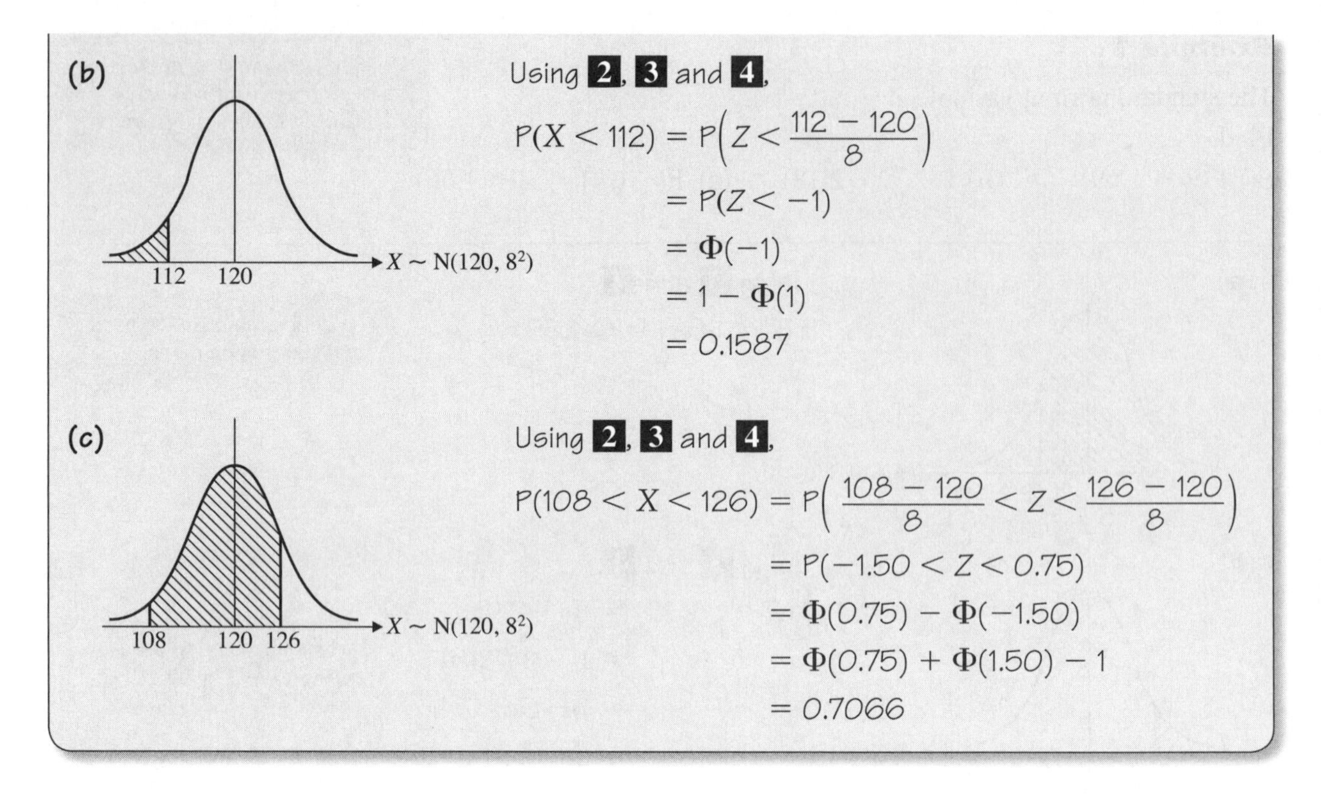

(b) Using **2**, **3** and **4**,

$$P(X < 112) = P\left(Z < \frac{112 - 120}{8}\right)$$
$$= P(Z < -1)$$
$$= \Phi(-1)$$
$$= 1 - \Phi(1)$$
$$= 0.1587$$

(c) Using **2**, **3** and **4**,

$$P(108 < X < 126) = P\left(\frac{108 - 120}{8} < Z < \frac{126 - 120}{8}\right)$$
$$= P(-1.50 < Z < 0.75)$$
$$= \Phi(0.75) - \Phi(-1.50)$$
$$= \Phi(0.75) + \Phi(1.50) - 1$$
$$= 0.7066$$

Example 3

The random variable X is normally distributed with mean 50 and standard deviation 10. Find the value of x such that:

(a) $P(X < x) = 0.2709$ **(b)** $P(X > x) = 0.0351$

(a)

0.2709

x 50 $X \sim N(50, 10^2)$

Using **2**, **3** and **4**,

$$P(X < x) = 0.2709$$
$$\therefore \quad P\left(\frac{X - 50}{10} < \frac{x - 50}{10}\right) = 0.2709$$
$$\therefore \quad P\left(Z < \frac{x - 50}{10}\right) = 0.2709$$
$$\Phi\left(\frac{x - 50}{10}\right) = 0.2709$$
$$\therefore \quad \Phi\left\{-\left(\frac{x - 50}{10}\right)\right\} = 1 - 0.2709 = 0.7291$$
$$\therefore \quad -\left(\frac{x - 50}{10}\right) = 0.61$$
$$\therefore \quad \frac{50 - x}{10} = 0.61$$
$$\therefore \quad x = 43.9$$

The value of x must be less than 50 since $P(X < 50) = 0.5$. Hence x is to the left of the mean and the z value corresponding to x will be negative.

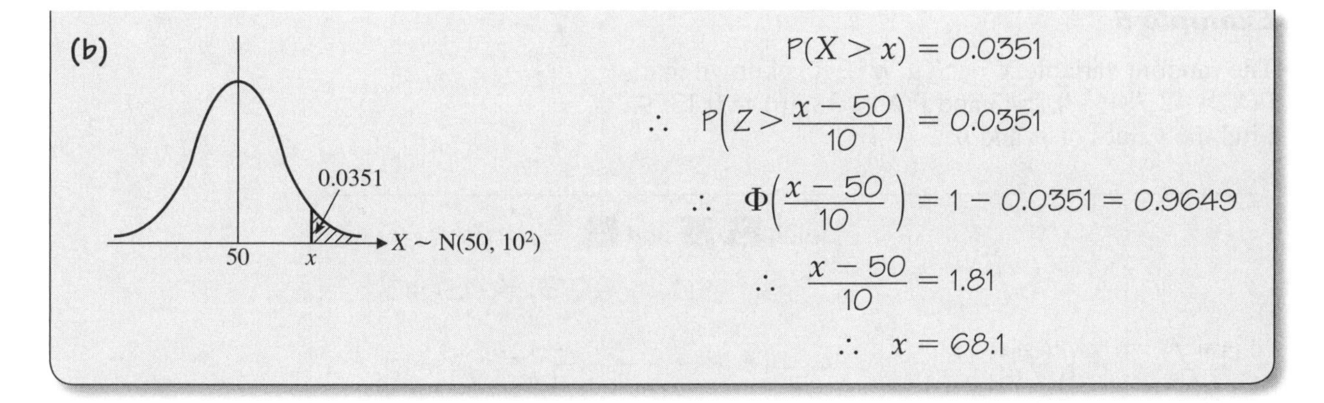

(b)

$$P(X > x) = 0.0351$$

$$\therefore \quad P\left(Z > \frac{x - 50}{10}\right) = 0.0351$$

$$\therefore \quad \Phi\left(\frac{x - 50}{10}\right) = 1 - 0.0351 = 0.9649$$

$$\therefore \quad \frac{x - 50}{10} = 1.81$$

$$\therefore \quad x = 68.1$$

Example 4

The weight of beans dispensed into a tin of beans has a normal distribution with mean 230 g and standard deviation 12 g. The weights of the empty tins into which the beans are dispensed are also normally distributed with mean 60 g and standard deviation 3 g. The weight of a tin is independent of the weight of the beans in it. Find the probability that a tin selected at random weighs less than 62 g and contains less than 247 g of beans.

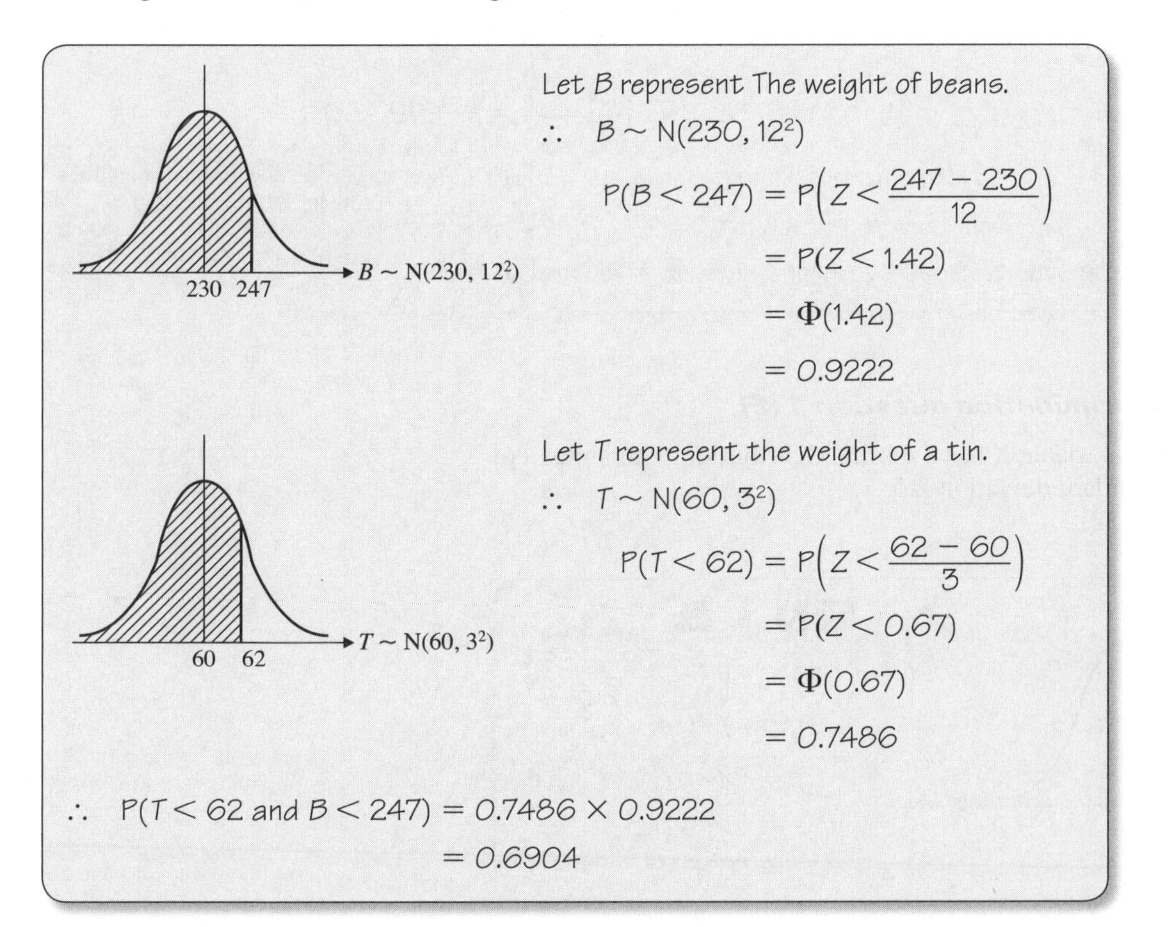

Let B represent The weight of beans.

$\therefore \quad B \sim N(230, 12^2)$

$$P(B < 247) = P\left(Z < \frac{247 - 230}{12}\right) = P(Z < 1.42) = \Phi(1.42) = 0.9222$$

Let T represent the weight of a tin.

$\therefore \quad T \sim N(60, 3^2)$

$$P(T < 62) = P\left(Z < \frac{62 - 60}{3}\right) = P(Z < 0.67) = \Phi(0.67) = 0.7486$$

$$\therefore \quad P(T < 62 \text{ and } B < 247) = 0.7486 \times 0.9222 = 0.6904$$

Example 5

The random variable $X \sim N(\mu, \sigma^2)$. It is known that $P(X > 48.78) = 0.2643$ and $P(X < 38.46) = 0.1379$.
Find the values of μ and σ.

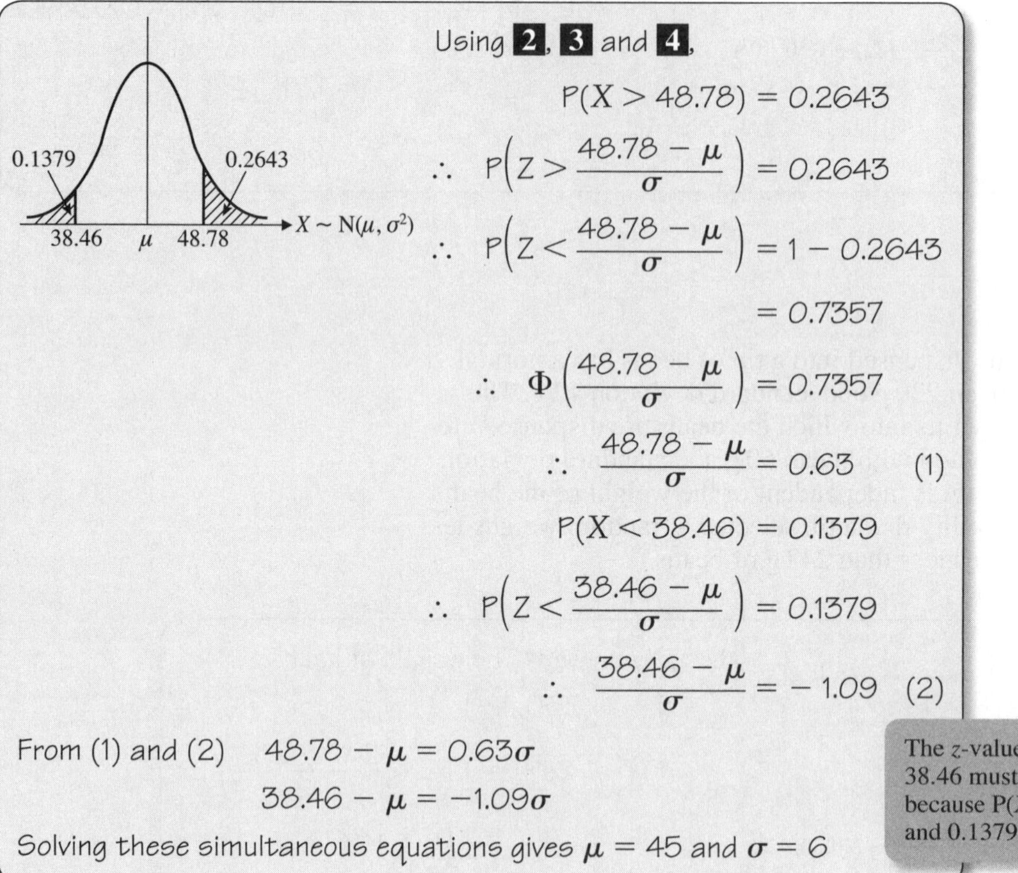

Using **2**, **3** and **4**,

$$P(X > 48.78) = 0.2643$$

$$\therefore \quad P\left(Z > \frac{48.78 - \mu}{\sigma}\right) = 0.2643$$

$$\therefore \quad P\left(Z < \frac{48.78 - \mu}{\sigma}\right) = 1 - 0.2643$$

$$= 0.7357$$

$$\therefore \quad \Phi\left(\frac{48.78 - \mu}{\sigma}\right) = 0.7357$$

$$\therefore \quad \frac{48.78 - \mu}{\sigma} = 0.63 \qquad (1)$$

$$P(X < 38.46) = 0.1379$$

$$\therefore \quad P\left(Z < \frac{38.46 - \mu}{\sigma}\right) = 0.1379$$

$$\therefore \quad \frac{38.46 - \mu}{\sigma} = -1.09 \qquad (2)$$

From (1) and (2) $\quad 48.78 - \mu = 0.63\sigma$

$$38.46 - \mu = -1.09\sigma$$

Solving these simultaneous equations gives $\mu = 45$ and $\sigma = 6$

The z-value is negative since 38.46 must be less than μ because $P(X < \mu) = 0.5000$ and $0.1379 < 0.5000$

Worked examination question 1 [E]

The random variable X has a normal distribution with a mean of 36 and a standard deviation 2.5.
Find $P(X < 32)$.

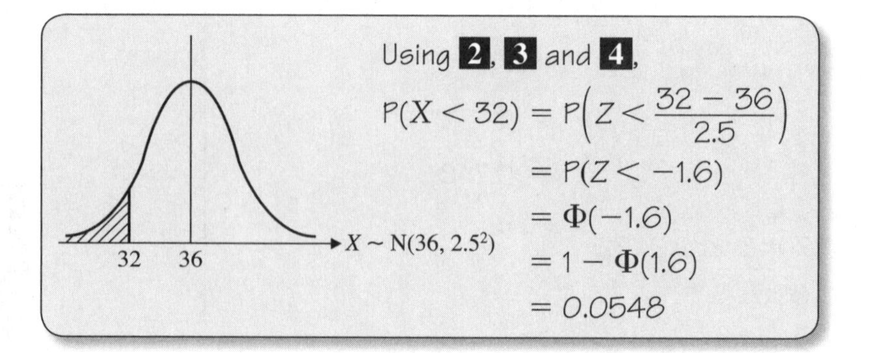

Using **2**, **3** and **4**,

$$P(X < 32) = P\left(Z < \frac{32 - 36}{2.5}\right)$$

$$= P(Z < -1.6)$$

$$= \Phi(-1.6)$$

$$= 1 - \Phi(1.6)$$

$$= 0.0548$$

Worked examination question 2 [E]

The random variable X is normally distributed with mean 100 and variance 225. Find:

(a) $P(X > 128)$ **(b)** $P(76 < X < 124)$

(c) the value of x such that $P(X < x) = 0.15$

The distribution of heights of adults is often modelled by a normal distribution.

(d) Give a reason to support the use of the normal distribution when modelling height.

(e) Give a reason why such a model may be unrealistic.

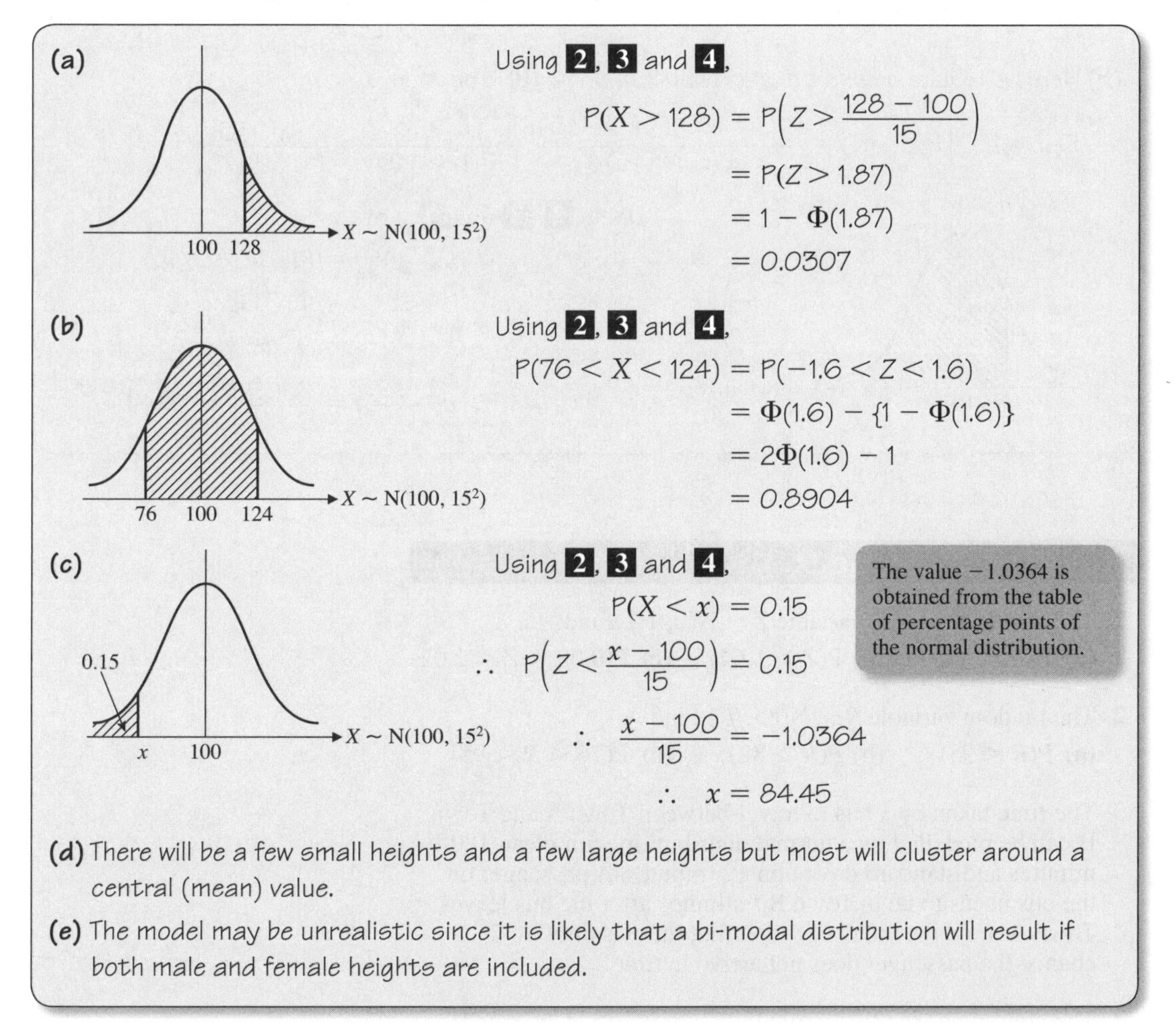

Worked examination question 3 [E]

The lifetime in hours of an electrical component has a normal distribution with mean 150 hours and standard deviation 8 hours. Find the probability that:

(a) a new component lasts at least 160 hours

(b) a component which has already operated for 145 hours will last at least another 15 hours.

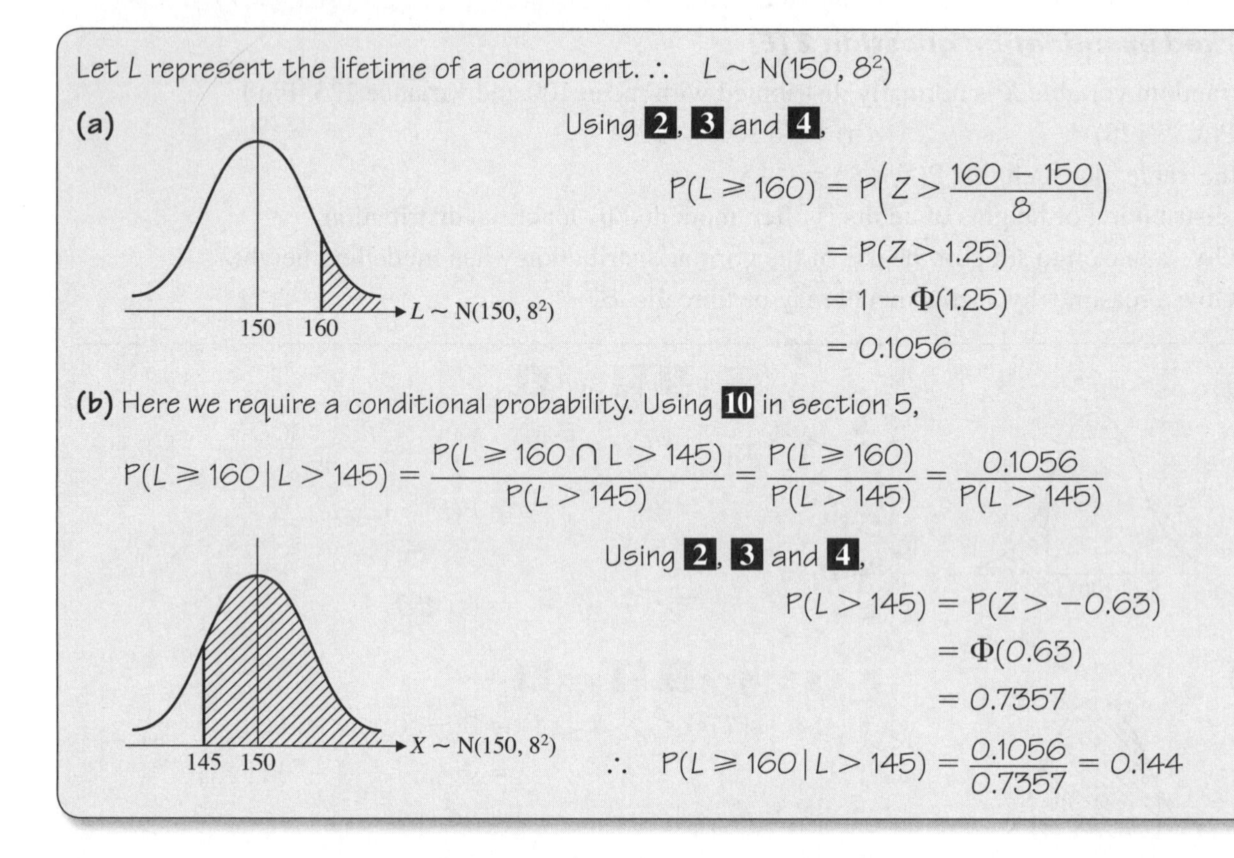

Let L represent the lifetime of a component. $\therefore \quad L \sim N(150, 8^2)$

(a)

Using **2**, **3** and **4**,

$$P(L \geqslant 160) = P\left(Z > \frac{160 - 150}{8}\right)$$
$$= P(Z > 1.25)$$
$$= 1 - \Phi(1.25)$$
$$= 0.1056$$

(b) Here we require a conditional probability. Using **10** in section 5,

$$P(L \geqslant 160 \mid L > 145) = \frac{P(L \geqslant 160 \cap L > 145)}{P(L > 145)} = \frac{P(L \geqslant 160)}{P(L > 145)} = \frac{0.1056}{P(L > 145)}$$

Using **2**, **3** and **4**,

$$P(L > 145) = P(Z > -0.63)$$
$$= \Phi(0.63)$$
$$= 0.7357$$

$$\therefore \quad P(L \geqslant 160 \mid L > 145) = \frac{0.1056}{0.7357} = 0.144$$

Revision exercise 9

1 The standard normal variable $Z \sim N(0, 1^2)$. Find:
(a) $P(Z < 1.19)$ **(b)** $P(Z > 1.62)$ **(c)** $P(0.08 < Z < 2.02)$

2 The random variable $R \sim N(85, 7^2)$ Find:
(a) $P(R < 93)$ **(b)** $P(R > 82)$ **(c)** $P(76 < R < 95)$

3 The time taken by a bus to travel between Town A and Town B can be modelled by a normal distribution with mean 180 minutes and standard deviation 12 minutes. A passenger on the bus needs to be in Town B t minutes after the bus leaves Town A. Find the value of t such that there is only a 1% chance the passenger does not arrive in time.

4 The continuous random variable X is normally distributed with mean μ ($\mu > 0$) and standard deviation σ. Given that $\mu = 2\sigma^2$ and that $P(X > 57) = 0.0808$, find $P(X < 48)$.

5 Pieces of wood are cut to a specified length L cm such that $L \sim N(\mu, \sigma^2)$. It is known that 1% of pieces are longer than 300.12 cm and that 0.5% are shorter than 299.87 cm. Find the value of μ and σ. Give your answer to 3 significant figures.

Examination style paper

Attempt all questions **Time 90 minutes**

In calculations you are advised to show all the steps in your working, giving your answer at each stage. Critical values from the Statistical Tables should be **quoted in full**. The answer to each part of a question which requires the use of tables or a calculator should be given to **three significant figures**, unless otherwise specified.

1 **(a)** Explain briefly what you understand by a statistical model.

(b) Write down a suitable distribution to model:

(i) the weight of a coffee jar

(ii) the suit of a playing card selected at random from a pack. **(4)**

2 The number of bags of crisps sold per day in a bar was recorded over a two-week period. The results are shown below.

20, 15, 10, 30, 33, 40, 5, 11, 13, 20, 25, 42, 31, 17

(a) Find the median and quartiles of these data.

An outlier is an observation that falls either
$1.5 \times$ (inter quartile range) above the upper quartile or
$1.5 \times$ (inter quartile range) below the lower quartile.

(b) Determine whether or not any items of data are outliers.

(c) Comment on the skewness of the distribution. **(8)**

3 Ten randomly selected children each sat a Mathematics test and an English test. Their scores in Mathematics (x) and in English (y) were recorded and summary statistics were calculated.

$\sum xy = 5877$, $\sum x = 288$, $\sum y = 224$, $S_{xx} = 1227.6$, $S_{yy} = 542.4$

(a) Calculate the product moment correlation coefficient between x and y.

(b) Give an interpretation of your coefficient.

The scores were then scaled by adding 10 marks to each Mathematics score and 8 marks to each English score.

(c) State, giving a reason, whether the coefficient would increase, decrease or stay the same. **(7)**

4 The table summarises the weights of a random sample of 200 livestock.

Weight (kg)	130–139	140–149	150–159	160–169
Number of livestock	3	20	47	71

Weight (kg)	170–179	180–189	190–199
Number of livestock	39	16	4

For these data $\sum fx = 32\,770$, $\sum fx^2 = 5\,398\,580$, $Q_1 = 155.24$, $Q_3 = 171.81$ where x represents the mid-point of each group and f represents the corresponding frequency.

(a) Calculate an estimate of the mean and an estimate of the standard deviation of the weights of the population of livestock from which this sample was taken.

(b) Find the median, Q_2, of this distribution.

One measure of skewness is given by

$$\frac{3\,(\text{mean} - \text{median})}{\text{standard deviation}}$$

(c) Evaluate this measure for the above distribution.

(d) Give a reason why a normal distribution might be suitable to model these data. **(11)**

5 Records show that the length, in mm, of a particular type of building block can be modelled by L where $L \sim \mathrm{N}(450, 3^2)$.

Find the probability that a randomly chosen block will be:

(a) longer than 455 mm

(b) between 444 and 452 mm long.

(c) Find the probability that of three randomly chosen blocks at least one will be longer than 455 mm.

(d) Determine the value of k such that 95% of the blocks will have their length in the range $(450 \pm k)$ mm. **(14)**

6 The discrete random variable X is defined as

$$\mathrm{P}(X = x) = \frac{1}{2kx} \qquad x = 1, 2, 3, 4.$$

(a) Find the value of k.

(b) Write down the probability distribution of X.

Find:

(c) $\mathrm{E}(X)$

(d) $\mathrm{Var}(X)$

(e) $\mathrm{E}(3X - 4)$

(f) $\mathrm{Var}(2X + 5)$ **(17)**

7 Pieces made by a certain lathe are subject to three kinds of defects, A, B, C. A sample of 1000 pieces was inspected and yielded the following results.

3.1% had a type A defect,
3.7% had a type B defect,
4.2% had a type C defect,
1.1% had both type A and type B defects,
1.3% had both type B and type C defects,
1.0% had both type A and type C defects,
0.6% had all three types of defect.

(a) Represent these data on a Venn diagram.

(b) What percentage had none of these defects?

(c) Calculate how many pieces had at least one defect.

(d) Find the percentage that had not more than one defect.

A piece is selected at random from this sample.

(e) Given that it had only one defect, find the probability that it was a type A defect. **(17)**

Answers

Revision exercise 1

1 Devise a model and use it to make predictions. Collect data from the real world and compare with predictions by testing statistically. If necessary, refine the model and try again.

2 Height of people, Battery life, Breaking strain of cables, and there are many other alternatives.

Revision exercise 2

1 (a)

Defects	*Tally*	*Frequency*	*Cum. freq.*
0	|||	3	3
1	~~||||~~	5	8
2	||||	4	12
3	||||	4	16
4	|||	3	19
5	|	1	20
6	||||	4	24
7	|||	3	27
8	||	2	29
9	|	1	30

(b) 10 batches have 6 or more defective mugs and so the manager will not need to buy from another manufacturer.

2 **(a)** 13 **(b)** 13.5 **(c)** 13.6 **(d)** mean

Revision exercise 3

1 **(a)** 10 **(b)** 10 **(c)** 13 **(d)** 14 **(e)** 10.3

2 **(a)** 172.2, 175.8, 178.4
(b) 183.9 **(c)** 175.55

3 **(b)** −0.29 **(c)** 175.55

4 **(a)** 20.25 **(b)** 3.455

5 5.812

Revision exercise 4

1 **(a)** 7, 55 **(b)** 57, 64

2 **(a)** Less
(b) **(i)** 17 mins **(ii)** 37 mins.

3 **(a)** Median 33, IQR 24
(b) **(d)**

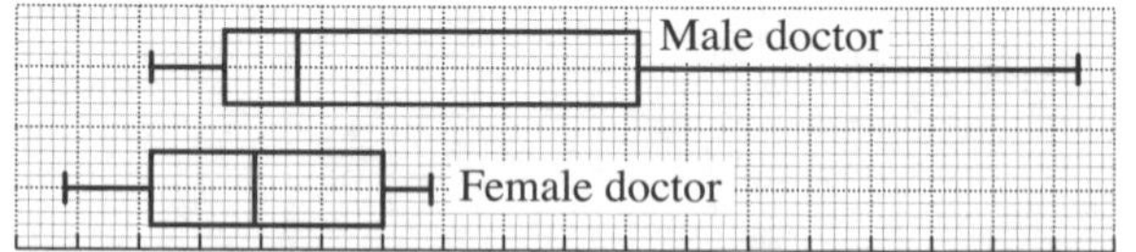

(c) 41.2, 20.7
(e) Median male > Median female
IQR male > IQR female
Range male > Range female
Male +ve skew; Female slight +ve/ almost symmetrical

4 **(a)** Number of CDs 3|1 means 31

3	1 4	(2)
4	3 3 5 7	(4)
5	1 1 2 2 2 5 6 8	(8)
6	1 1 2 4 5 6 6 7 8 8	(10)
7	1 2 3 5 5 5 6 6 8 9	(10)
8	3 3 6 8 8 9	(6)

(b) 6
(c) Few students with small number of CDs – majority had 60 or more.

5 **(a)** Variable – weight is a continuous variable.

(b)

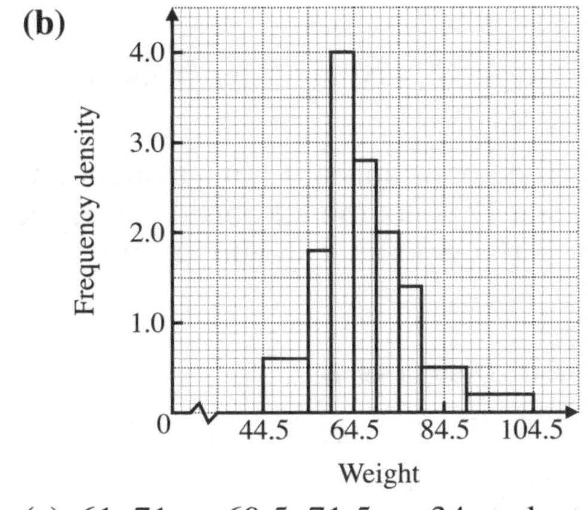

(c) 61–71 ⇒ 60.5–71.5 ⇒ 34 students

Revision exercise 5

1 **(a)** Independent **(b)** Mutually exclusive

2 **(a)** 0.58 **(b)** 0.6

3 **(a)** $\frac{3}{35}$ **(b)** $\frac{2}{25}$

4 **(a)**

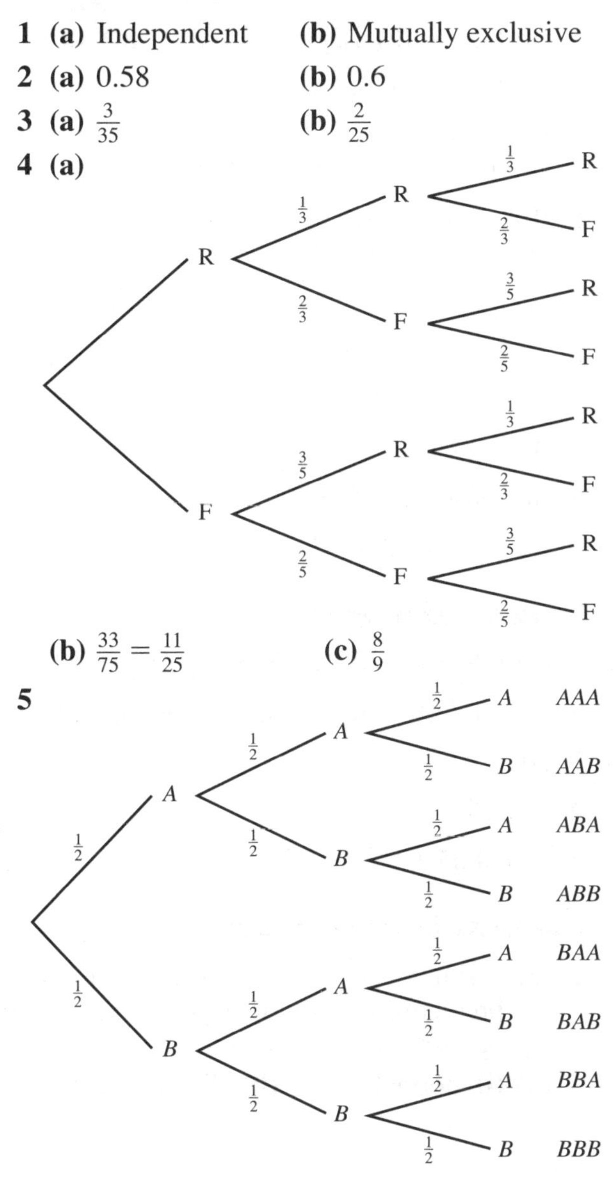

(b) $\frac{33}{75} = \frac{11}{25}$ **(c)** $\frac{8}{9}$

5

(a) $\frac{1}{8}$ **(b)** $\frac{3}{8}$ **(c)** $\frac{1}{2}$

(d) The system is fair to each side since P(A) = P(B)

6 **(a)** $\frac{1}{2}, \frac{11}{12}$ **(b)** $\frac{6}{7}$ **(c)** $\frac{4}{5}$

Revision exercise 6

1 **(a)** $S_{xx} = 14, S_{yy} = 58.83, S_{xy} = 23$

(b) 0.8014

(c) This is a reasonably high positive correlation as might be expected with poetry which has to stick to a metre.

2 **(a)** 0.9406

(b) There appears to be a high positive correlation. The greater the yield the more labels are needed.

3 **(a)** 0.4277

(b) The correlation is not high but it is positive. In the circumstances of war this is to be expected.

Revision exercise 7

1 **(a)** $y = 14.5 + 1.02t$

(b) a = temperature of milk when brought into room

(c) 19.1, 35

(d) $t = 4.5$ is closer to the mean

(e)

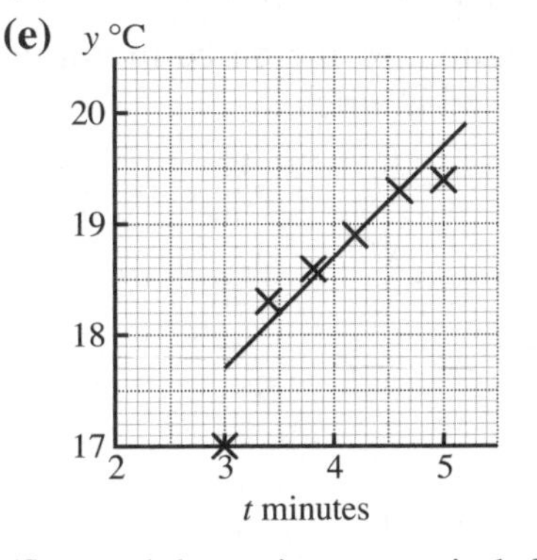

(f) model ⇒ t increases indefinitely so could exceed room temperature.

(g) Model is OK for small values of t but a curve that is asymptotic as $t \to \infty$ would be better.

2 (a)

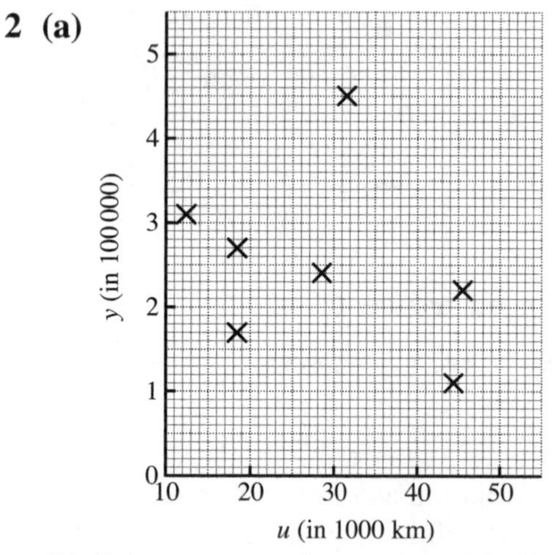

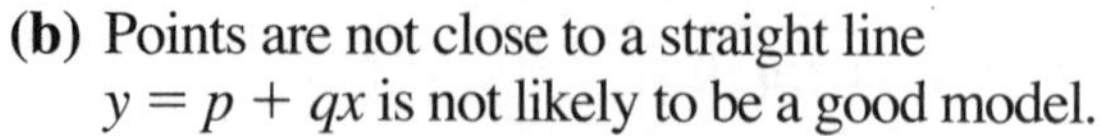

(b) Points are not close to a straight line $y = p + qx$ is not likely to be a good model.

(c) and **(f)**

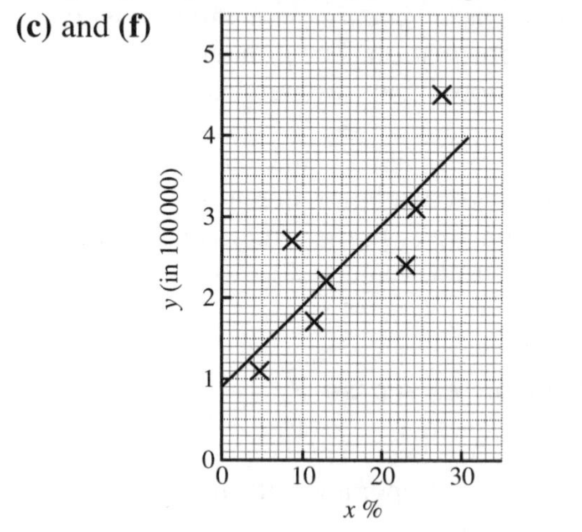

(d) $y = 0.956 + 0.098x$

(e) Number of injuries if no motorways

(g) The line appears a reasonable fit.

(h) +ve. Gradient ⇒ increased motorway ⇒ increased injuries.

3 (a)

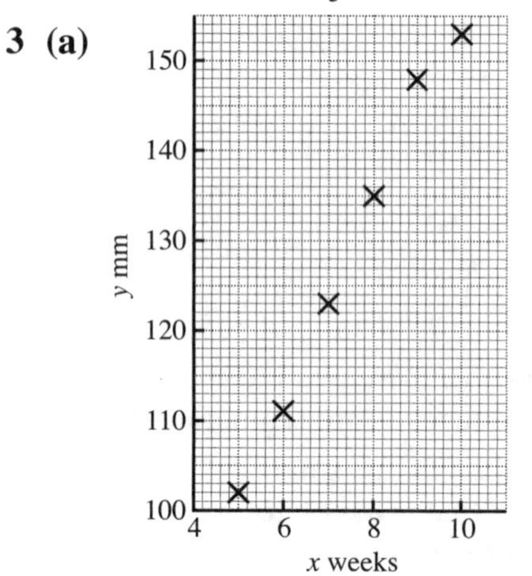

(b) $y = 47.67 + 10.8x$

(c) Height of seedling when planted.

(d) 177.27

(e) The points appear to fit a straight line and 12 is not too far outside the range 5–10 weeks so it should be fairly reliable.

Revision exercise 8

1 **(a)** 0.6 **(b)** 0.7 **(c)** 3.0 **(d)** 1.2 **(e)** 17.4

2 **(a)** Discrete uniform

(b) $F(x) = \frac{x}{6}, \quad x = 1, 2, 3, 4, 5, 6$

(c) **(i)** $\frac{1}{2}$ **(ii)** $\frac{1}{2}$ **(d)** **(i)** $3\frac{1}{2}$ **(ii)** $\frac{35}{12}$

3 **(a)** 0.5 **(b)** 1.3 **(c)** 3.3 **(d)** 1.61

4 **(a)** 0.1, 0.2, 0.5, 1.0.

(b)

x:	0	1	2	3
p(x):	0.1	0.1	0.3	0.5

5 **(b)**

x:	0	1	2	3	4
p(x):	$\frac{9}{36}$	$\frac{12}{36}$	$\frac{10}{36}$	$\frac{4}{36}$	$\frac{1}{36}$

(c) $1\frac{1}{3}$

6 **(a)**

x:	-5	-3	-1	2	4	6
p(x):	$\frac{1}{6}$	$\frac{1}{6}$	$\frac{1}{6}$	$\frac{1}{6}$	$\frac{1}{6}$	$\frac{1}{6}$

(b) $\frac{1}{2}$

(d) Growing. $E(X) > 0$

(e) $\left(\frac{1}{6}\right)^4 = 0.000\,77$

Revision exercise 9

1 **(a)** 0.8830 **(b)** 0.0526 **(c)** 0.4464

2 **(a)** 0.8729 **(b)** 0.6664 **(c)** 0.8251

3 207.9

4 0.3446

5 300, 0.0510

Examination style paper

1 **(a)** A statistical process to describe or make predictions about the expected behaviour of a real-world problem.

(b) **(i)** normal **(ii)** discrete uniform.

(a) $Q_2 = 20$, $Q_1 = 13$, $Q_3 = 31$
(b) No outliers
(c) positive skew $Q_3 > Q_2 > Q_2 - Q_1$

(a) -0.704
(b) Those who did well in Maths did not do well in English and vice versa.
(c) Coefficient would stay the same since PMCC is an index and is not affected by linear transformations of the variables.

4 (a) 163.85, 12.12
(b) 163.73
(c) -0.030
(d) Skewness is very close to zero. For a normal distriution skewness is zero. Thus a normal distribution might be a suitable model.

5 (a) 0.0475 **(b)** 0.7258
(c) 0.136 **(d)** 5.88

6 (a) $\frac{25}{24}$
(b)

x	1	2	3	4
$P(X = x)$	$\frac{12}{25}$	$\frac{6}{25}$	$\frac{4}{25}$	$\frac{3}{25}$

(c) 1.92 **(d)** 1.1136
(e) 1.76 **(f)** 4.4544

7 (a)

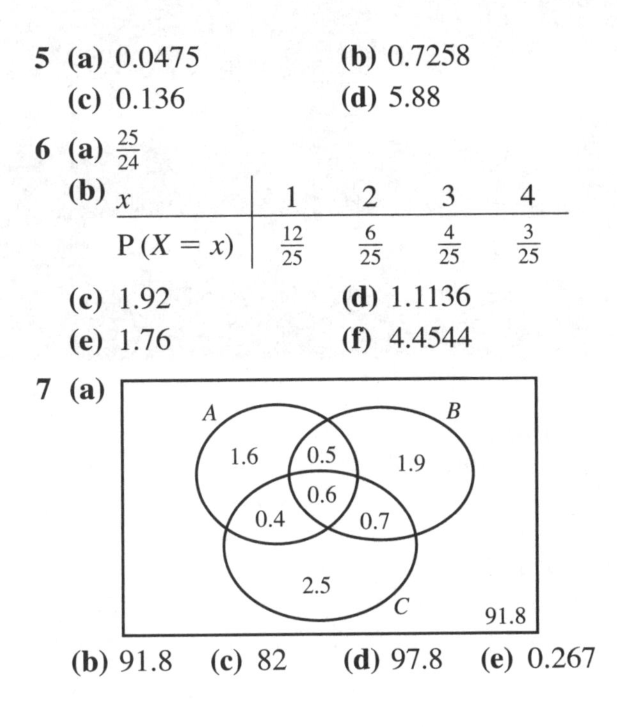

(b) 91.8 **(c)** 82 **(d)** 97.8 **(e)** 0.267

Test yourself answers

Chapter 2

1 (b) −0.528, 381.86

Chapter 3

1 (a) 13 (b) 12.5 (c) 10 (d) 11 (e) 11.8

2 (a) 10 (b) 9, 10, 12 (c) 11 (d) 10.36

3 (a) 374.54, 379.63, 387.97 (b) 402.71 (c) 381.86

4 (a) 30 (b) 16

5 2.635, 1.285

6 10.703, 1.486

Chapter 4

1 (a) 0 | 6 means 6 Passengers (b) Only collected data for one day.

0	6 9	(2)
1	0 0 1 1 6 7 7 9	(8)
2	0 1 3 3 4 7 8 9	(8)
3	0 2 4 5 5 7 7	(7)
4	0 1 2 7 9	(5)

2 (a) 19.5, 29.5; 24.5; 10 (b) 0.6, 0.9, 2.4, 3.4, 0.5, 0.1

3 (a) 2.9, 6.0 (b) 1.5, 3.8

4 (a) 67 (b) 61, 69.5, 80

(c)

40 50 60 70 80 90 100

Number of CDs

(d) 70.28

5 (a) 8, 56 (b)

0 10 20 30 40 50 60 70

6 **(a)** 30.1, 25.4, 36.9 **(b)** positive skewness

(c) One would expect a few long consultations but most would be short $\Rightarrow$ positive skewness

(d) 32.65, 160.67

(e) The median and quartiles should be used, due to the skewness of the distribution.

(f) and **(g)**

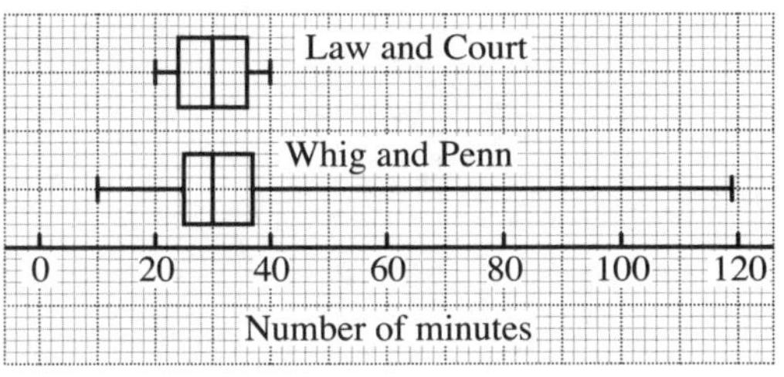

(h) Any sensible comments relating to location and spread
e.g. same median; Law and Court symmetrical, Whig and Penn positively skewed; range of Whig and Penn much greater than range of Law and Court.

Chapter 5

1 **(a)** $\frac{1}{4}$ **(b)** $\frac{3}{8}$ **(c)** $\frac{11}{16}$

2 $\frac{1}{10}$

3

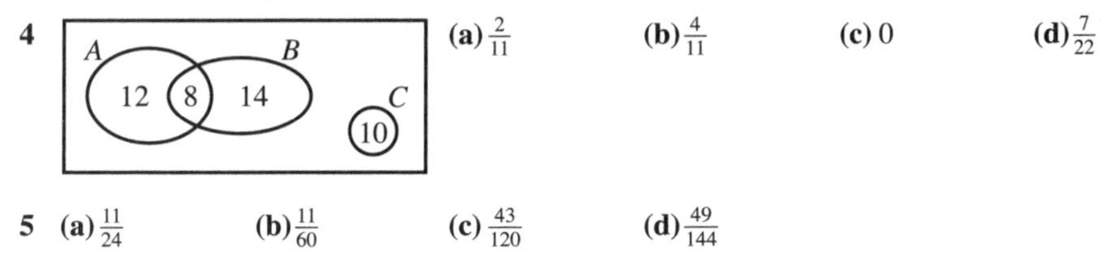

(a) $\frac{1}{2}, \frac{3}{13}, \frac{1}{13}$ **(b)** $P(A \cap C) = 0$ so mutually exclusive

(c) $P(B \cap C) = \frac{6}{52}$ and $P(B) \times P(C) = \frac{26}{52} \times \frac{12}{52} = \frac{6}{52}$ so they are independent.

4

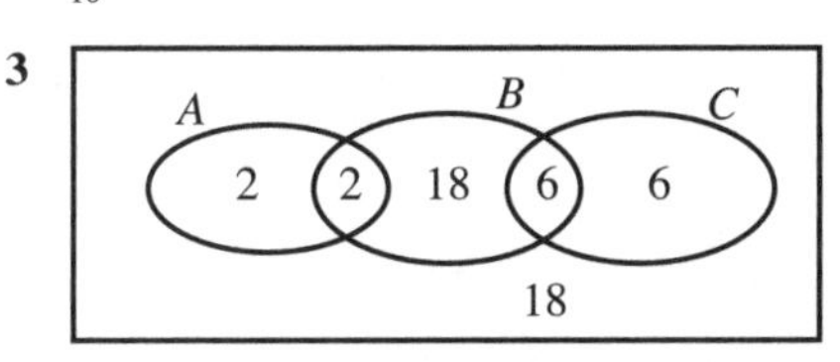

(a) $\frac{2}{11}$ **(b)** $\frac{4}{11}$ **(c)** 0 **(d)** $\frac{7}{22}$

5 **(a)** $\frac{11}{24}$ **(b)** $\frac{11}{60}$ **(c)** $\frac{43}{120}$ **(d)** $\frac{49}{144}$

Chapter 6

1 **(a)** $r = -1$ **(b)** $r = 0$ **(c)** $r = 0.8$

2 **(a)** $S_{xx} = 1428.1$, $S_{yy} = 736$, $S_{xy} = 469$

(b) $r = 0.457$

(c) 0.457 is not a high positive correlation but it is positive so it gives some indication. The aptitude test was done before training had taken place and the weekly sales measured the sales person's potential after the training course so these include the influence of the training course as well as the recruit's raw potential.

3 **(b)** 0.954

(c) High positive correlation. This supports the department's claims.

Chapter 7

1

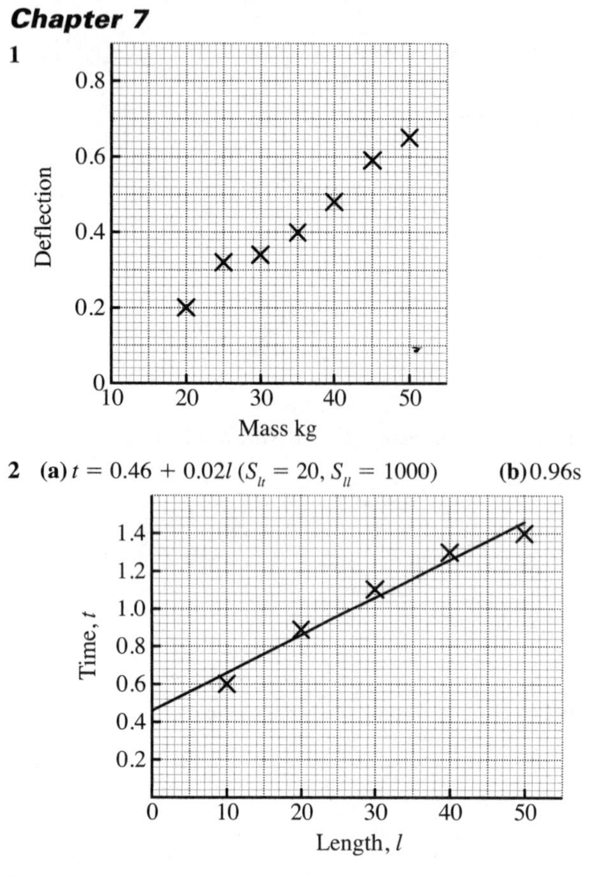

2 **(a)** $t = 0.46 + 0.02l$ ($S_{lt} = 20$, $S_{ll} = 1000$) **(b)** 0.96s

3 **(a)** a represents the number of loaves consumed when there are no people attending so it should be 0.
(b) b is the average number of loaves consumed per person.

Chapter 8

1 $E(X) = -1$; $Var(X) = 1.2$

2 **(a)** $F(3) = 1$ **(c)**

x:	1	2	3
$P(X = x)$:	$\frac{1}{4}$	$\frac{3}{8}$	$\frac{3}{8}$

3 **(a)**

s:	2	3	4	5	6	7	8	9
$P(S = s)$:	$\frac{1}{20}$	$\frac{1}{10}$	$\frac{3}{20}$	$\frac{1}{5}$	$\frac{1}{5}$	$\frac{3}{20}$	$\frac{1}{10}$	$\frac{1}{20}$

(b) $E(S) = 5\frac{1}{2}$ **(c)** $Var(S) = \frac{13}{4} = 3\frac{1}{4}$

4 **(a)** $E(2X + 2) = 12$ **(b)** $Var(2X + 2) = 16$ **(c)** $E(X^2) = 29$

5 **(a)** $p(d) = \frac{1}{8}$, $d = 1, 2, , \ldots, 7, 8$. **(b)** $P(7 \leqslant D \leqslant 8) = \frac{1}{4}$; A fair charge would be 25p.

Chapter 9

1 **(a)** 0.9236 **(b)** 0.0384 **(c)** 0.8427

2 **(a)** 0.3156 **(b)** 0.0823 **(c)** 0.6708

3 $d = 617.9\text{ km} \Rightarrow 617\text{ km}$

4 **(a)** 5 **(b)** 0.9502

5 $\mu = 3468.24 \Rightarrow 139$ bags; $\sigma = 139.71 \Rightarrow 6$ bags